OZARK PLANTS

Trees, Shrubs, Wildflowers and Grasses
of the Ozark Mountains of
Arkansas, Missouri and Oklahoma

STEVE CHADDE

Ozark Plants
Trees, Shrubs, Wildflowers and Grasses of the Ozark Mountains
of Arkansas, Missouri and Oklahoma

AN ORCHARD INNOVATIONS FIELD GUIDE
Printed in the United States of America

ISBN 978-1951682552

The author may be contacted by email at: steve@orchardinnovations.com
VERSION 1.0, 01/22/2022

Contents

Preface

A version of this work was first published in 1969 by the U.S. Forest Service as *Ozark Range and Wildlife Plants* (authors Hewlette S. Crawford, Clair L. Kucera, and John H. Ehrenreich). The goal of the original publication was to provide a field guide for forest ecologists, wildlife biologists, and range management specialists in the Ozark region for identifying plants and describing their value for wildlife and livestock. The current work expands on the original, updates plant nomenclature (the scientific names), and adds color photographs for nearly all species described. Wildlife and livestock importance and use is largely derived from the original work, and readers interested in the studies cited, both published and unpublished, can refer to the 1969 publication (available at: https://catalog.hathitrust.org/Record/011392688).

In this book, plants are first grouped into one of the following sections: **trees and shrubs**, **wildflowers**, and **grasses and grasslike plants**. Within each section, plants are listed alphabetically by their scientific genus or species name. The **family index** on page 293 lists all species within their respective families.

Plants included in this work were chosen as they are (1) widespread and abundant in the Ozarks, (2) have some food value for livestock or wildlife, or (3) have distinct ecological or economic values. For each species, general distribution and site preferences, characteristics, and importance are described.

Scientific nomenclature largely follows (with few exceptions), the currently accepted names as listed by the *Integrated Taxonomic Information System* (ITIS, www.itis.gov). In cases where an older name is still widely used or there are conflicting reports of current names, these are provided as synonyms for the species.

Keys to the genera (page 254) for plants in both the summer and winter seasons, and a **glossary** (page 288) are provided. Some training in taxonomy will be helpful in using the keys but is not essential.

Illustrations are from a number of sources: the author's collection, the public domain, and photographers who have made their work available for commercial use under a creative commons license. The author gratefully acknowledges the contribution these many individuals have made in recording our natural heritage.

Introduction

The flora of the Ozark Mountain region is abundant and diverse for several reasons, including a long presence uninterrupted by glaciation, varied site and soil conditions, and a midcontinental position. Many species with distributional centers outside of the Ozarks reach their geographic limits here. The range of certain plants from southwestern United States extends to the Ozarks, typically in dry-soil environments such as the glades. On moist north-facing slopes, plants from the northern and eastern United States may be found. Some typically southern species occur in special habitats, including many parts of the Boston Mountain Plateau. Overall, nearly 3,300 vascular plant species are known from the region, within approximately 1066 genera and 186 families (Biota of North America Program, 2021); this book includes descriptions for over 500 species.

The values of these many species for livestock, wildlife (primarily game species) and other uses, described in this handbook, are based largely on published studies, unpublished records, and observations. Information for nongame species is scarce. Unless stated otherwise, the forage evaluations refer to the season when the plants are used most and to the area in which they are most common. The food values for wildlife are so variable that we have not attempted to rank the food species by palatability.

DESCRIPTION OF THE OZARK REGION

The Ozark region is a plateau and mountain area of more than 40,000 square miles in southern Missouri, northern Arkansas, and the northeastern corner of Oklahoma. Also sometimes considered a part of the region is a small area, sometimes designated as the Ozark Hills, in the unglaciated section of southern Illinois, where elevations rise above 1,000 feet.

The principal part of the region is bounded on the north and east by the Missouri and Mississippi River valleys. Loessal hills between these valleys and the Ozark region form a transitional area of hardwood forest and farmland. From east-central Missouri, the Ozark plateau extends southwest to the Boston Mountains in Arkansas. These mountains, the southernmost part of the Ozarks, run east to west for approximately 200 miles along the north edge of the broad Arkansas River valley into northeastern Oklahoma. The western edge of the Ozarks, extending from Oklahoma northeastward across Missouri, is in transition with the prairie and savanna region, and topographically forms the least distinct boundary.

The Boston Mountains are higher than the rest of the Ozarks, with many elevations exceeding 2,300 feet. The highest point in the Missouri Ozarks is Tom Sauk Mountain, 1,772 feet, in the eastern part of the state. The average elevation in the prairie transition of the western Ozarks is about 1,000 feet, compared to only 400 feet along the Mississippi lowlands.

The Ozark hill and valley pattern was cut through an ancient plateau. Steep slopes with elevational differences of 400 to 800 feet or more are frequent. But some sections have a more undulating terrain of broad valleys and low hills. Igneous rocks of pre-Cambrian origin, standing out as high points or knobs in the eastern Missouri Ozarks, are among the oldest in North America. Surrounding this core in a concentric pattern are sedimentary rocks of sandstone and dolomitic and calcitic limestone, primarily of Ordovician and Mississippian ages.

SOILS

The most extensive soils of the Ozarks are those formed by the weathering of the underlying rock formations. In addition, there are alluvial deposits along streams, colluvial soils on the lower slopes, and limited areas of loessal materials. The last type occurs primarily as caps on ridgetops and flat terrain. Soils of the upper slopes and ridges are generally infertile and stony; chert fragments are often abundant on the surface of the ground.

Soils of the Clarksville series cover wide areas in the central and west-central Ozarks. These soils, derived from cherty dolomitic limestone and sandstones of the Jefferson City and Rubidoux formations, have very low fertility and moderate to rapid permeability. The Fullerton soils, similar to the Clarksville soils and also widespread, are generally more fertile, with moderate profile development. In the eastern Missouri Ozarks, soils of the Ashe series, formed from igneous rocks, are shallow and have low fertility and moderate permeability.

The "cedar glade" region of the White River drainage has extensive soils classified in the Gasconade series, formed from siliceous dolomitic outcrops of the Cotter formation. Glade soils are dark, generally alkaline, and moderately fertile. They average only a few inches deep and dry out quickly. Because of their low water-holding capacity and the rapid runoff on sloping terrain, vegetation tends to be sparse.

The forested areas scattered throughout the glades occur on soils with deeper profiles and greater water retention capability. These light-colored forest soils of the Corydon, Nixa, Bodine, and Baxter series are cherty and generally infertile, and were developed from limestone rock. The Baxter soils can retain more moisture, which is available to plants, than can the Bodine and Corydon soils. The Bodine are strongly leached and acidic, and the Corydon are shallow and calcareous. Nixa soils occur on ridge-tops, are characterized by a hardpan and slow permeability, and generally support a forest cover of poor quality.

Scattered throughout the region are areas of flat terrain called "post oak flats" with silty clay subsoils, hardpans, and slow drainage. These light-colored forest soils of the Lebanon series have low fertility and are derived from loess overlying limestone residuum. The Dickson soils are a sloping phase associated with the Lebanon soils.

Soils in the Boston Mountains are derived chiefly from Pennsylvanian sandstones and shales. These include the extensive Rarden (formerly Wedington), Hector, Jefferson, and Pottsville soils of varying depth. The Pottsville soils have shallow profiles with poor soil-water relationships, and support chiefly a forest and native grass vegetation. Rarden soils have plastic clay subsoils derived from shaly sandstone and generally are poor sites for timber. Most of the Hector soils are wooded but are not as well suited for timber production as the Jefferson soils.

CLIMATE

The Ozark region has long warm summers and mild to moderately cold winters. Mean annual temperatures vary between 53° and 60° F. Average July temperatures range from 76° to 78° in the north to about 80° in the southeast. January averages range from 30° to 32° along the northern and western borders to nearly 40° in Stone County, Arkansas, in the southeastern Ozarks. Growing seasons vary from 170 or 180 days in the northwest to about 215 days in the southeast.

Precipitation ranges from 50 to 55 inches in the southeast to about 40 inches in the northwest. Growing-season rainfall varies from about 25 to 28 inches throughout the Ozarks. During the dormant season, however, rainfall ranges from only about 15 or 16 inches in the northwest to 25 inches or more in the southeast.

VEGETATION

The Ozark region is at the western edge of the eastern hardwood forest and forms a transition with prairie and savanna vegetation in northeastern Oklahoma, extreme southeastern Kansas, and western and central Missouri. To the south, the Ozarks border the shortleaf pine-hardwood forests of the Ouachita Mountains.

Oaks of varying density and species are the prevailing forest vegetation. In certain areas, mainly in the eastern and southern sections, shortleaf pine (*Pinus echinata*) is an associated species. Pines are most prevalent on south and west slopes with droughty soils of sandstone and abundant chert, where competition from hardwoods is not as critical as on north or east slopes or on deeper soils. Clarksville soils from the Rubidoux formation in Missouri and Rarden-Hector-Pottsville soils farther south are characteristic pine-bearing soils. Oak or oak-pine forests are commonly interspersed with natural gladelike openings supporting varying quantities of herbaceous plants.

Throughout the Ozarks little bluestem (*Schizachyrium scoparium*) and big bluestem (*Andropogon gerardi*) afford varying amounts of forage and ground cover. Species of *Panicum* and *Dichanthelium* are especially numerous. Indian grass (*Sorghastrum nutans*), purpletop (*Tridens flavus*), and several species of paspalum (*Paspalum* spp.) are common. Along roadsides and in forest stands of low densities or where openings occur, grasses together with various broad-leaved herbaceous plants may be relatively abundant. Under open forest conditions, wild lespedeza (*Lespedeza* spp.) and tick trefoil (*Desmodium* spp.) are common legumes in the ground cover. These legumes, together with big and little bluestem, Indian grass, and switch grass (*Panicum virgatum*) are the most important native plants for livestock forage in the Ozarks.

Typical glade vegetation is most extensive in southwestern Missouri and northern Arkansas in the White River drainage area. Among principal grasses on exposed rocky slopes are big and little bluestem, side-oats grama (*Bouteloua curtipendula*), and "baldgrass" (*Sporobolus* spp.). Eastern redcedar (*Juniperus virginiana*) is a common tree in the glades and in some areas is increasing in density. Other tree or shrub species in the glades are smoke-tree (*Cotinus obovatus*), persimmon (*Diospyros virginiana*), and aromatic sumac (*Rhus aromatica*).

Glades and other shallow-soil sites are a significant source of uncultivated native forage, and total several million acres. Glade lands alone occupy more than 500,000 acres in the Missouri Ozarks and about an equal acreage in the Arkansas Ozarks. Over time, woody plants tend to invade and dominate these sites, thereby reducing both the productivity and the diversity of grass and wildflower species. The timber-producing potential of these sites is extremely low; thus, they are best maintained for grazing and for wildlife food and cover.

OZARK MOUNTAIN ECOREGIONS

The general area covered by this book includes ecoregions 8.4.5 **OZARK HIGHLANDS** (a large portion of southern Missouri and northern Arkansas, and small portions of northeastern Oklahoma and southeastern Kansas) and 8.4.6 **BOSTON MOUNTAINS** (north of the Arkansas Valley [8.4.7] and south of the Ozark Highlands [8.4.5] in northwestern Arkansas and northeastern Oklahoma) (source: US EPA).

Vegetation of the OZARK HIGHLANDS: Oak-hickory and oak-hickory-pine forest are typical. Some savannas and tallgrass prairies were once common in the vegetation mosaic. Post oak, blackjack oak, black oak, white oak, hickories, shortleaf pine, little bluestem, Indiangrass, big bluestem, eastern red cedar glades.

Terrain: More irregular physiography than adjacent regions, with the exception of the Boston Mountains (8.4.6) to the south. Mostly a dissected

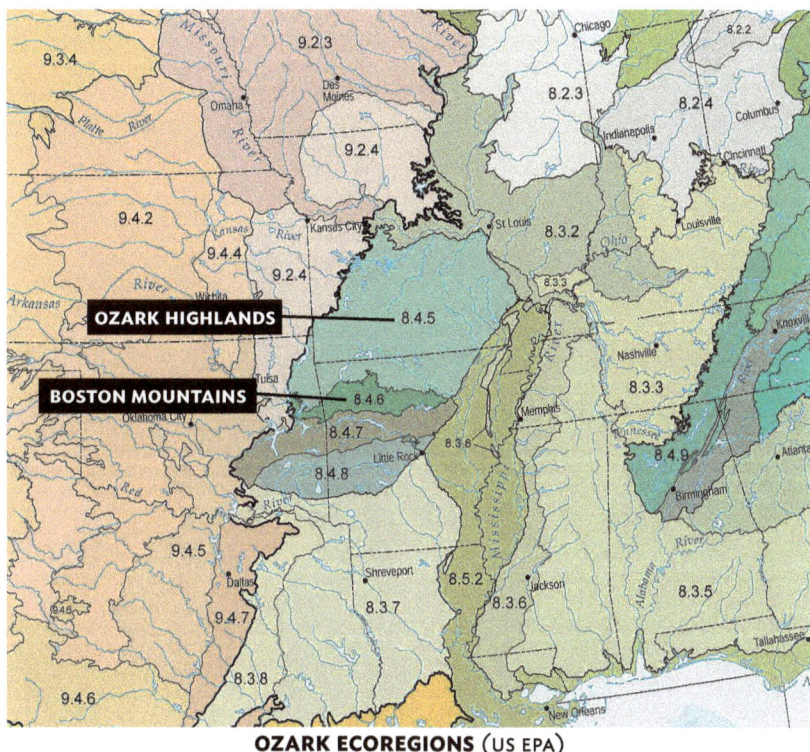

OZARK ECOREGIONS (US EPA)

limestone plateau, the region has karst features including caves, springs, and spring-fed streams. Some steep, rocky hills, and some gently rolling plains. Elevations range from 80 m to 560 m. Limestone, chert, sandstone, and shale are common, some small areas of igneous rocks in the east.

Vegetation of the BOSTON MOUNTAINS: Mostly oak-hickory forests. Red oak, white oak, post oak, blackjack oak, and hickories remain the dominant vegetation types in this region, although shortleaf pine and eastern red cedar are found in many of the lower areas and on some south- and west-facing slopes. Some mesophytic forests in ravines and on north-facing slopes with sugar maple, beech, red oak, white oak, basswood, and hickory.

Terrain: A deeply dissected mountainous plateau, in contrast to the nearby Ouachita Mountains (8.4.8) which comprises folded and faulted linear ridges. Elevations range from 65 m to 853 m. Geology is mostly Pennsylvanian-age sandstone, shale, and siltstone, in contrast to the limestone and dolomite of the adjacent Ozark Highlands (8.4.5).

TYPICAL OZARK FOREST. A mix of oaks and other hardwoods with scattered shortleaf pine (center).

Trees and Shrubs

This section treats plants with woody or partially woody stems and includes trees and most vines and shrubs. Woody plants are generally considered important wildlife-food producers but less important as a source of food for livestock. However, woody plants browsed by range livestock help supply adequate protein after the protein content of grasses declines in midsummer.

Stems of woody plants are not as important for wildlife food in the Ozarks as in other regions, probably because fruit-producing and herbaceous plants are available to supply palatable year-round food. Where these plants are crowded out by dense timber stands or where they have been reduced in abundance by localized overpopulations of deer, there is a greater dependence on woody stems for wildlife food.

POST OAK, *Quercus stellata.* A common oak throughout the Ozark region and the southeastern United States (see page 58).

Prairie Acacia

Description Low semi-shrubby perennial with slender pubescent stems; leaves alternate, pinnately compound with numerous fine leaflets about 1/16 in. (1½ mm.) wide or less; flowers minute, cream colored, in dense spherical clusters ½–¾ in. (¼–2 cm.) in diameter, developing in late spring and summer; pod flat, up to 3 in. (7½ cm.) long and about ⅜ in. (9–10 mm.) wide.

PRAIRIE ACACIA, *Acaciella angustissima*

Synonym *Acacia angustissima*
Distribution Throughout the southern and western parts of the Ozark range, north to southern Missouri.
Habitat Glades, prairies, and open woods; on dry rocky soils.
Importance This species is not abundant in the Ozarks and does not contribute greatly to the total diet of livestock or deer. However, it is highly nutritious, is eaten by all kinds of livestock during spring, summer, and fall, and will decrease under heavy grazing.

Although no wildlife utilization records were found for the Ozarks, the seeds of this species are eaten by doves and quail in the southwestern United States and probably are used in the Ozark region also.

Acer spp.

Maple, Boxelder

Description Forest trees, occasionally tall, with straight bole in favorable locations; leaves simple, opposite, palmately lobed, with long petioles, or pinnately compound, with several leaflets; flowers small, greenish or reddish, clustered, at ends of branches, appearing in spring; fruit a double samara, 2-seeded.

SUGAR MAPLE, *Acer saccharum*

BLACK MAPLE, *Acer nigrum*

SUGAR MAPLE, *fruit*

Distribution Widespread and sometimes common and abundant.
Habitat Upland areas or ravines, rich slopes, bottomlands, and floodplains.
Sugar maple, *Acer saccharum*, of ravines and coves, also on uplands, is identified by 5-lobed leaves, glabrous below, whereas the less common **black maple**, *Acer nigrum*, has more or less 3-lobed leaves, or the two basal lobes are less distinct than in sugar maple and are pubescent on the underside. In these two species, flowers appear with or after the leaves.

Two floodplain or bottomland trees are **silver maple**, *Acer saccharinum*, and **red maple**, *Acer rubrum*. The first species has deeply lobed leaves and oblong but sharp sinuses, and the latter shallower sinuses; the leaves of both are whitish or silvery on the underside. **Boxelder**, *Acer negundo*, is also a floodplain type, with compound leaves, the leaflets 3 or 5 and coarsely dentate on the margin, and with green twigs. In these three

Maple, Boxelder

BOX ELDER, *Acer negundo*

RED MAPLE, *Acer rubrum*

SILVER MAPLE, *Acer saccharinum*

floodplain species, the flowers generally appear before the leaves unfold, and the fruits mature much earlier than in sugar maple or black maple.

Importance Sugar maple provides food for ruffed grouse, especially during spring and summer. Buds and seeds of maples are eaten during spring and summer and to a lesser extent during autumn by squirrels. Sugar maple and red maple twigs and buds are browsed by deer during winter. Livestock make some use of these plants during spring and autumn.

Buckeye

Description Shrubs or small forest trees with opposite, palmately compound leaves and thick brown twigs; leaflets 5, elliptic to obovate-lanceolate, with straight parallel veins from the midrib and fine serrations on the margins; flowers yellowish or red in upright showy clusters, appearing in spring; fruit a leathery nutlike capsule with one or more large shiny seeds.

RED BUCKEYE, *Aesculus pavia*

OHIO BUCKEYE, *Aesculus glabra*
ABOVE, RIGHT

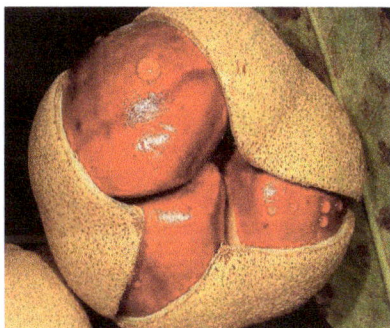

Distribution Widespread, or one species more restricted.
Habitat Low woods, ravines, and bottomlands; on moist soils.

Two principal species are the widespread and common **Ohio buckeye**, *Aesculus glabra*, with yellow flowers and dark-brown seeds; and the **red buckeye**, *Aesculus pavia*, mostly in more eastern parts of the Ozarks, and identified by red flowers and light-brown seeds.
Importance Cattle, sheep, and horses may browse the young tender shoots in spring if other forage is lacking. Later in the season livestock are usually not attracted to

the shoots. Hogs will eat the seeds, particularly in the fall, and squirrels occasionally eat them.

Cattle, sheep, horses, and hogs have been poisoned from eating buckeye. The toxic material is especially prevalent in young sprouts and seeds.

Humans may be affected by eating the seeds. Children especially should not have access to the attractive nutlike seeds. However, Native Americans reportedly roasted and ate the nuts with no apparent injury. Bee keepers claim the flowers have been poisonous to honey bees.

Downy Serviceberry, Juneberry, Shadbush

Description Small tree with smooth gray bark and elongate sharp-pointed buds; leaves simple, alternate, ovate to obovate, with finely serrate margins, densely pubescent when opening; flowers conspicuous, white, with 5 strap-shaped petals, appearing in spring; fruit reddish to purple, berrylike.

DOWNY SERVICEBERRY, *Amelanchier arborea*

Distribution Widely distributed.
Habitat Open woods, bluffs, and dry slopes, less common on low ground.
Importance The fruits of Juneberry are eaten by squirrels in June and July. Reports from other regions show that many songbirds also eat these fruits. After the buds swell in the spring, they are eaten by ruffed grouse. Deer occasionally browse this species on overpopulated ranges, and cattle occasionally nip at it throughout the year.

Humans frequently eat the berries, either raw or prepared in pies and pastries.

Pawpaw

Description Small understory tree, usually forming thickets; leaves simple, alternate, obovate and tapering to short petiole, large blade to 12 in. (30 cm.) long, with entire margins; flowers conspicuous, maroon or purplish, about 1 in. (2½ cm.) in diameter, appearing in spring; fruit soft, edible, oblong, 4–5 in. (10–12 cm.) long, with several seeds.

PAWPAW, *Asimina triloba*

Distribution Widely distributed.
Habitat Low rich woods and ravines; on deep soils.
Importance Since this tree is not abundant in the Ozark uplands, it does not provide much wildlife food. However, the "bananalike" fruits are eaten by squirrels from August through October, and by raccoons. Pawpaw is occasionally browsed by cattle in winter. This tree is usually found in bottoms where cattle concentrate and is browsed more because cattle are near the tree than because animals seek it.

When ripe, pawpaw fruits provide an interesting and different taste for humans.

Supplejack, Rattan Vine

Description Shrubby plant with twining branches and smooth greenish bark; leaves simple, alternate, oblong, 1–3 in. (2½–7½ cm.) long, glabrous, the venation prominent and evenly spaced, entire on the margin; flowers inconspicuous, greenish white, appearing in spring; fruit black, berrylike, with 2 seeds.

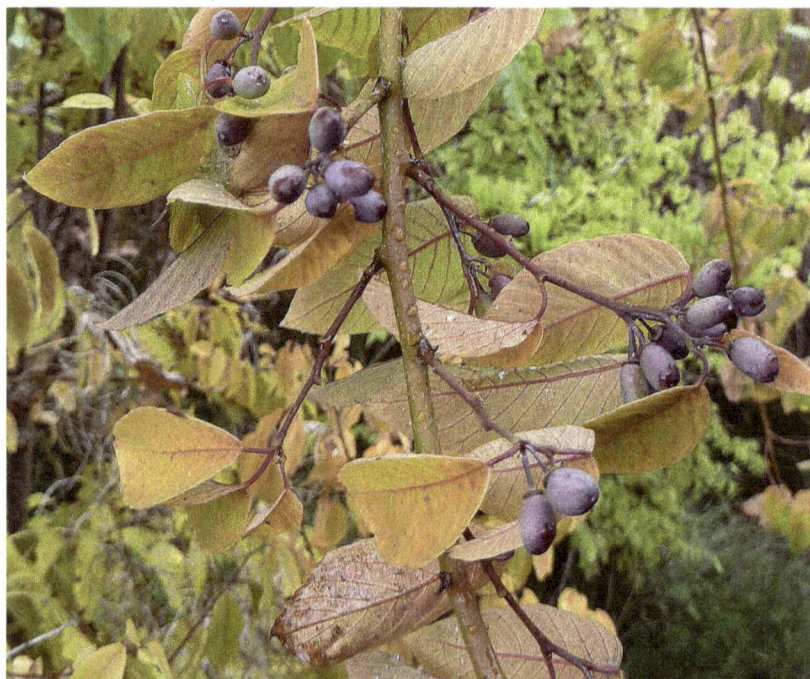

SUPPLEJACK, *Berchemia scandens*

Distribution Scattered through central and southern Ozarks.

Habitat Limestone glades, ravines, and stream bottoms.

In the Mississippi lowlands, supplejack occurs as a high climbing vine, and is quite different in habit from the Ozark type.

Importance Supplejack occurs sparingly and is not an important food for wildlife in the Ozarks. This plant is important for deer, raccoons, and songbirds in other regions where it is more abundant.

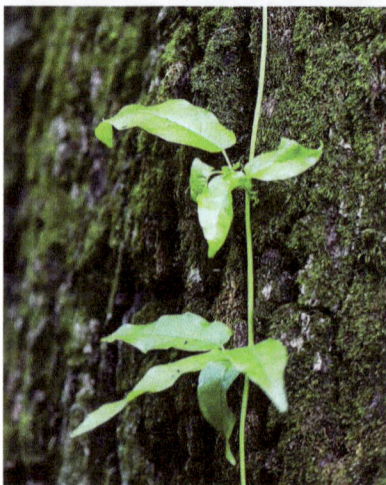

Beauty-berry, French Mulberry

Description Shrub to 6–7 ft. (2 m.) tall; leaves simple, opposite, ovate, with tapering base and pointed tip, and whitish beneath with minute teeth on the margins; flowers pinkish, numerous, in clusters at base of leaves, developing in late spring and summer; fruit lavender, smooth, berrylike, later becoming dried.

BEAUTY-BERRY, *Callicarpa americana*

Distribution Southern Ozarks north to Taney County, Mo.

Habitat Wooded slopes, under pine and oak; on dry to moist soils.

Importance This plant occurs very sparingly in the northern and central Ozarks and consequently is not an important food for wildlife except possibly in the southern Ozarks. In other regions where more abundant, beauty-berry is browsed by deer and its fruits are eaten by songbirds.

Traditionally, people would rub their clothing and skin with the leaves to ward off biting insects, such as ticks, deer flies, ants, and mosquitoes. Research by the USDA has confirmed the presence of several insect-repelling compounds in the plant.

Hickory

Description A large complex group of forest trees with alternate, pinnately compound leaves, stout twigs, and mostly large terminal buds; leaflets 5 or more, elliptic to obovate or lanceolate, with pointed tip and tapering base, particularly for terminal leaflet, and finely to coarsely serrate on the margins; staminate flowers in 3-branched catkins, the pistillate flowers in short spikes or clusters; nut globose or oblong with splitting husk.

PIGNUT HICKORY, *Carya glabra*

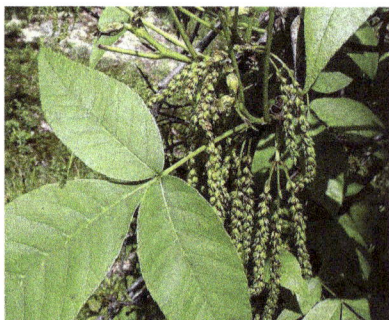

SHAGBARK HICKORY, *Carya ovata*
ABOVE, RIGHT

Distribution Widespread and sometimes common and abundant.
Habitat Dry upland woods, or on moist soils in ravines and bottomland.

Common species of upland sites include the **shagbark hickory**, *Carya ovata*, with large terminal buds, 5 leaflets, and light bark in loose strips; mockernut, Carya tomentosa, identified also by large terminal buds but with 7 or 9 leaflets, and dark tight bark; and **pignut hickory**, *Carya glabra*, with 5 or 7 leaflets, the bark becoming shaggy in older trees, but terminal buds smaller than in shagbark, only ½ in. (10–12 mm.) long or less.

An important hickory of dry woodlands is **black hickory**, *Carya texana*, with 5 or 7 leaflets, dark furrowed bark, and reddish pubescence on young twigs. **Bitternut hickory**, *Carya cordiformis*, of ravines and bottomlands on deep soils, has 7 or 9 leaflets and conspicuous yellowish buds. Other hickories also occur in the region.

Carya spp.
Hickory

BLACK HICKORY, *Carya texana*

BITTERNUT HICKORY, *Carya cordiformis*

Importance Hickory nuts are one of the most preferred foods of fox and gray squirrels. Squirrels also eat hickory buds. Deer seldom browse hickories when the range is in good condition, but may browse succulent new growth on depleted range. Hickory foliage is browsed by livestock only when other food is scarce. Hogs compete with wildlife for the nuts.

Ozark Chinkapin

Description Small to medium-size tree; leaves simple, alternate, narrow-elliptic to oblong, side veins of leaf blade mostly parallel; conspicuously dentate staminate flowers in elongate catkins, somewhat arching or ascending, with numerous stamens, pistillate flowers in a short ovoid cluster, developing in late spring; fruit a nut, enclosed in a spiny bur.

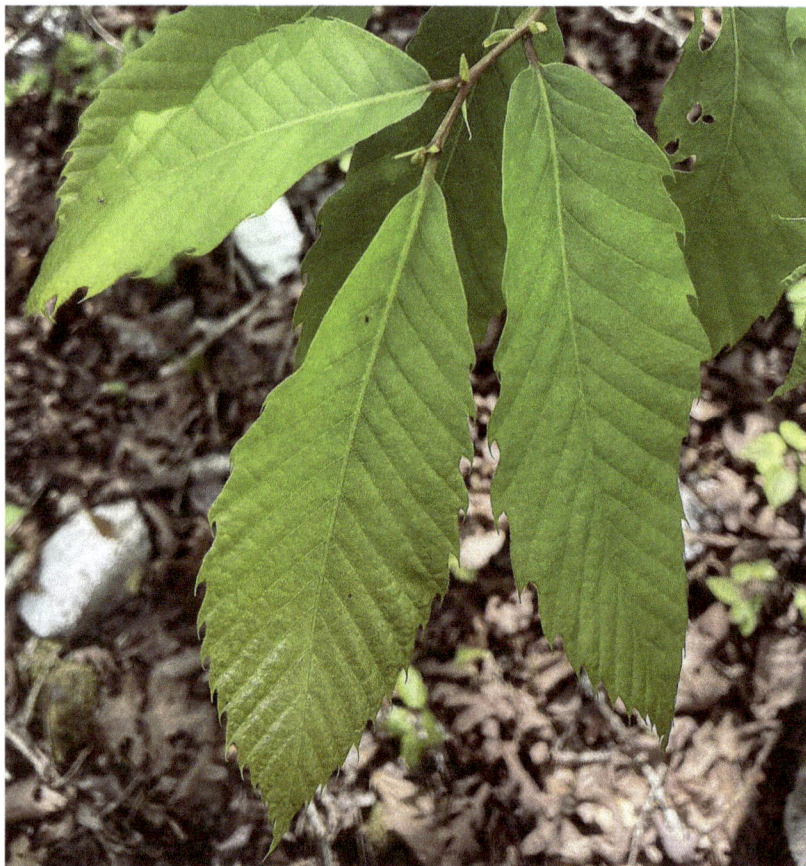

OZARK CHINKAPIN, *Castanea ozarkensis*

Distribution Arkansas north to southern Missouri Ozarks.
Habitat Upland slopes and ridges; on cherty or sandy, mostly acid soils.

Importance Squirrels eat the ripened nuts, but the fruit often falls before it ripens. Mature trees are scarce and Ozark chinkapin cannot be considered an important food.

New Jersey Tea, Redroot

Description Low shrubs usually less than 3 ft. (90 cm.) high; leaves simple, alternate, ovate or oblong to lanceolate, somewhat hairy on underside, with minute teeth on the margin; flowers small, white, in terminal clusters or from axils of leaves, appearing in spring; fruit a dry splitting capsule with 3 seeds.

REDROOT
Ceanothus herbaceus

NEW JERSEY TEA
Ceanothus americanus

Distribution Widely distributed. **Habitat** Prairies, glades, and open woods; on generally dry soils.

New Jersey tea, *Ceanothus americanus*, is distinguished by ovate or oblong leaves mostly more than 1 in. (2.5 cm.) wide and somewhat elongated flower clusters.

Redroot, *Ceanothus herbaceus*, has narrower leaves and more flat-topped (umbel-like) flower clusters. **Importance** Deer browse New Jersey tea repeatedly during the growing season and to a lesser extent during the dormant season. This plant is considered fair to good forage for cattle, sheep, and goats during the summer. Turkeys occasionally eat the seeds.

These small shrubs have a deep taproot and are capable of fixing nitrogen in the soil.

New Jersey tea derived its name from its use as a tea substitute during the Revolutionary War.

Bittersweet

Description Woody, twining, vinelike shrub; leaves simple, alternate, broadly oblong, the apex pointed, with finely scalloped margin; flowers small, greenish, in loose racemes, appearing in late spring; fruit a splitting orange capsule with scarlet "seeds" persisting in the fall and winter.

BITTERSWEET, *Celastrus scandens*

Distribution Widespread and sometimes common and abundant.

Habitat Glades, open woods, thickets, along fence rows, and in waste ground.

Importance The persistent bittersweet fruits are one of the more important grouse winter foods. Squirrels may use the fruits as an emergency food; deer also occasionally eat them. The fruits and leaves are said to be toxic to livestock; however, few poisonings have been reported.

The bright fruits provide a colorful contrast to a sometimes drab winter scene.

Celtis spp.

Hackberry, Sugarberry

Description Small to large forest trees with warty bark; leaves simple, alternate, ovate to lanceolate, noticeably netveined on lower side, toothed or almost entire on the margin; flowers small, greenish, appearing in spring; fruit small, berrylike, mostly dark purplish.

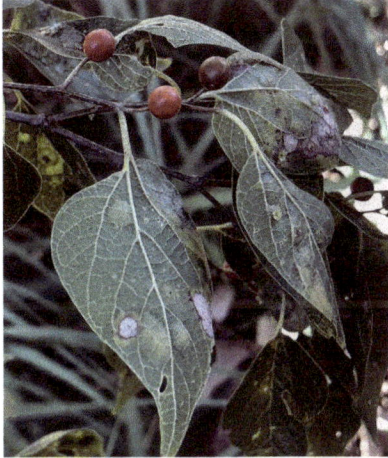

SUGARBERRY, *Celtis laevigata*

HACKBERRY, *Celtis occidentalis* →

DWARF HACKBERRY
Celtis tenuifolia

Distribution Widely distributed. Habitat Rocky upland woods, glades, pastures, to bottomlands and floodplain; on alluvial soils.

Hackberry, *Celtis occidentalis*, and the less prevalent **sugarberry**, *Celtis laevigata*, are generally lowland trees. The first species has toothed oblique leaves, the latter species more or less entire or remotely toothed leaves with more even tapering bases.

Dwarf hackberry, *Celtis tenuifolia*, is a small tree of limestone glades and rocky woods, with broadly ovate short-pointed leaves, which are oblique at the base.

Importance Hackberry fruits as a preferred fall and winter food of

turkeys. Squirrels occasionally eat the fruits and, to a lesser extent, the buds and bark. Raccoons utilize the fruits as a principal food. Cattle will occasionally browse hackberry and sugarberry heavily, especially during winter on poor ranges. Hackberry makes up a small part of the diet of pheasants, waterfowl, quail, and ruffed grouse.

Buttonbush

Description Coarse shrub; leaves simple, opposite, oblong-elliptic, pointed, shiny green, entire on the margins; flowers small, whitish, in ball-shaped clusters from a stalk, appearing in summer; fruit in spherical heads, several-seeded.

BUTTONBUSH, *Cephalanthus occidentalis*

Distribution Widely distributed.
Habitat Swamps, borders of ponds and streams, and in flatwoods.
Importance Small quantities of but-tonbush seeds are eaten by water-fowl, especially wood ducks, and occasionally by pheasants.

Cercis canadensis

Redbud

Description Small tree of the legume family with dark bark; leaves simple, alternate, rounded or heart-shaped, with entire margins; flowers rose colored, in small clusters on the branches, appearing in spring before the leaves; pod about 3 in. (8 cm.) long, with several hard seeds.

REDBUD, *Cercis canadensis*

Distribution Widely distributed.
Habitat Glades, bluffs, and open woods.
Importance The seeds, buds, and bark of redbud are occasionally eaten by squirrels, probably as emergency foods. Deer eat redbud on ranges where more desirable foods are lacking, and cattle browse it occasionally on winter range.

Dogwood

Description Shrubs and understory trees; leaves simple, usually opposite, with conspicuous venation, entire on the margin; true flowers small, greenish or whitish, in rounded clusters; in flowering dogwood, there are also 4 conspicuous white petal-like bracts surrounding the flower cluster; fruits berrylike, white, blue, or bright-red.

FLOWERING DOGWOOD
Cornus florida

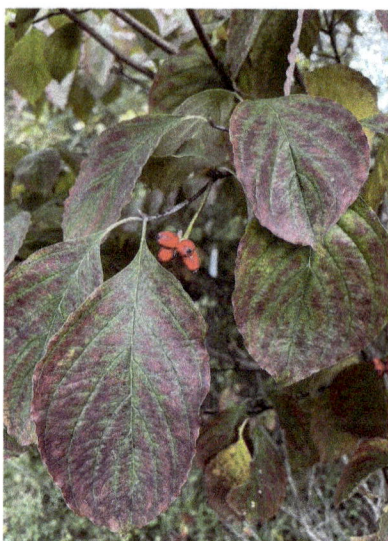

FLOWERING DOGWOOD
Cornus florida

Distribution Widely distributed.
Habitat Upland woods, glades, and waste ground to bottomlands and moist sites. Flowering dogwood is poor forage for cattle, possibly because of the bitter quininelike taste of the bark and leaves.

Flowering dogwood, *Cornus florida*, is a small understory tree in oak and pine woods with ovate leaves, pale beneath, and a bright-red shiny fruit.

Gray dogwood, *Cornus racemosa*, a shrub with light-green foliage and white berries, occurs in low woods and along stream bottoms.

The **roughleaf dogwood**, *Cornus drummondii*, also has white fruits and occurs in a wide range of sites including barrens and glades.

Cornus alternifolia in the eastern part of the range is a shrub with dark-green foliage and blue fruits and is the only species with alternate leaves.

Other species of varying distribution and site requirements also occur in the region.

Dogwood

ROUGHLEAF DOGWOOD, *Cornus drummondii*

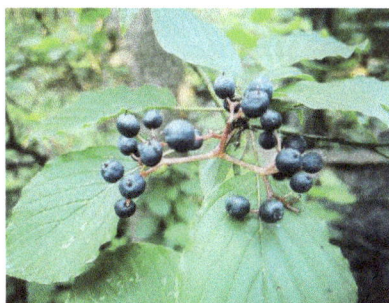

Cornus alternifolia

GRAY DOGWOOD, *Cornus racemosa*

Importance Dogwoods are not an especially important deer browse plant on ranges in good condition, but where ranges are overpopulated or have suffered past abuse, dogwoods form an important part of a deer's diet.

The abundant flowering dogwood fruits are eaten throughout much of the year by turkeys but are most important from September through January. Dogwoods are considered as preferred turkey food. Grouse, pheasants, and, to a lesser extent, wood ducks, quail, prairie chickens, and squirrels also eat dogwood fruits. Grouse sometimes eat the buds and flowers during spring.

Hazelnut

Description Coarse shrub or small tree; leaves simple, alternate, ovate to oblong, with pointed apex and rounded or cordate base, coarsely serrate on the margin; staminate flowers in pendulous spikelike catkins, the pistillate flowers in minute reddish clusters from buds, appearing in early spring; nut smooth, spherical, enclosed in a leafy involucre.

HAZELNUT, *Corylus americana*

staminate flowers

Distribution Widespread and sometimes common.

Habitat Thickets, open woods, and waste ground.

Importance Hazelnut catkins provide important winter food for grouse, and the nuts are an important squirrel food from September to January. Deer browse the twigs and catkins, primarily during winter.

Cotinus obovatus

American Smoketree

Description Small tree with usually a full rounded crown and spreading branches; inner wood yellow; leaves simple, alternate, rounded-elliptic, glaucous or bluish green, entire on the margin; flowers small, in delicate panicles, in midspring; fruit small, about 1/5 in. (5 mm.) long, the fruiting sprays giving a smoky appearance.

AMERICAN SMOKETREE
Cotinus obovatus

Distribution Southwestern Missouri and adjacent Arkansas and Oklahoma.
Habitat Limestone glades and barrens.
Importance These common glade plants are browsed lightly by cattle in early summer.

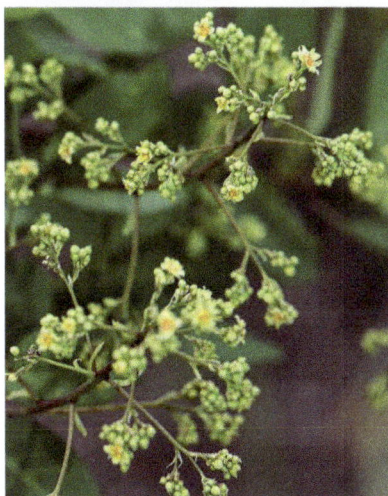

Hawthorn

Description A large complex group of shrubs or small trees, with generally thorny twigs and branches, but some species thornless; leaves simple, alternate, of various shapes, degrees of lobing, and serration; flowers conspicuous, with 5 white petals, generally clustered, appearing in midspring; fruit reddish, small, applelike, generally less than ½ in. (12–15 mm.) in diameter.

DOWNY HAWTHORN
Crataegus mollis

COCKSPUR HAWTHORN
Crataegus crus-galli

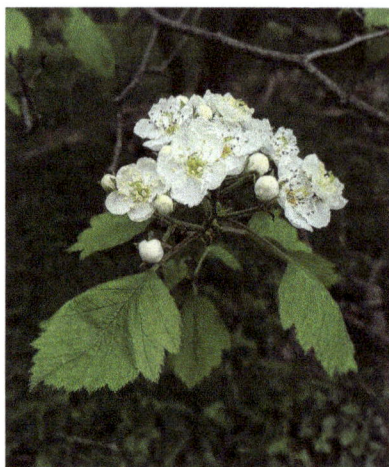

Distribution Widespread, some species common.
Habitat Prairies, glades, open woods, pastures, fields, and waste ground.

Two of the more common species are **cockspur hawthorn**, *Crataegus crus-galli*, with glossy obovate leaves, broadly rounded at apex and tapering to the base, and long thorns, and **downy hawthorn**, *Crataegus mollis*, with soft broadly ovate lobed leaves somewhat cordate or truncate at base, and generally thorny branches.

Numerous other species also occur in the range, but their taxonomy is complicated.
Importance Hawthorn fruits provide a small amount of food for wood ducks, pheasants, ruffed grouse, and squirrels. Deer occasionally eat the fruit and leaves. Hawthorn is occasionally browsed by cattle on winter range when native range grasses are scarce.

Diospyros virginiana

Persimmon

Description Small to medium-size tree, often forming thickets, or sometimes large forest trees on deep soils; bark thick, blocky, dark colored; leaves simple, alternate, ovate-oblong, 4 in. (10 cm.) long or more, entire on the margins; flowers pale green or yellow, 4-parted, appearing in spring; fruit yellowish, plumlike but usually several-seeded, about 1 in. (2½ cm.) in diameter, with astringent taste before maturity. A less common dark-fruited variety also occurs.

PERSIMMON, *Diospros virginiana*

Distribution Widespread and sometimes common and abundant.
Habitat Old fields, pastures, glades, also stream bottoms and low ground.
Importance Persimmon fruits are usually moderately abundant in the Ozarks and provide good fall and winter food for raccoons, red foxes, coyotes, deer, squirrels, and, to a lesser extent, quail. Many birds probably eat these fruits, but the pulp is difficult to identify in crop analyses. Hogs also relish the fruit, but persimmon is of little value to other livestock, and is considered a nuisance because it sprouts readily and invades abused range.

Euonymus spp.   CELASTRACEAE, BITTERSWEET FAMILY

Burning-bush, Eastern Wahoo, Strawberry Bush

Description Coarse shrub or small tree, or creeping or trailing, with green twigs somewhat ridged or 4-angled; leaves simple, opposite, either oblong-lanceolate with long tapering apex or obovate, finely serrate; flowers small, purplish, in loose clusters, appearing in spring; capsule lobed, splitting, with bright-red "seed," persisting through winter.

BURNING-BUSH
Euonymous atropurpureus

STRAWBERRY-BUSH
Euonymous obovatus

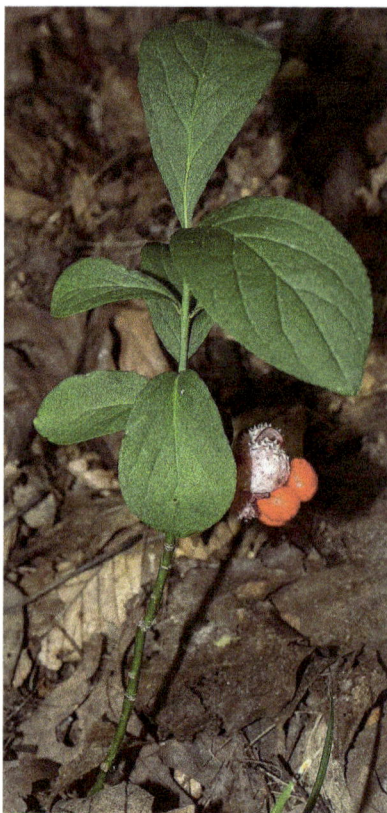

Distribution Widely distributed.
Habitat Wooded slopes, ravines, streambanks, and alluvial bottoms.
  Burning-bush or **Eastern wahoo**, *Euonymus atropurpureus*, is an upright shrub or small tree common throughout the Ozarks.
Strawberry-bush, *Euonymus obovatus*, is a trailing plant found in southern and eastern parts of the range.
Importance Burning-bush is occasionally browsed by deer as is strawberry bush, but neither is abundant enough to be important in the total diet. Burning-bush has purgative properties in the leaves and fruit and is seldom browsed by livestock when more desirable forage is available.

34 | TREES AND SHRUBS

Fraxinus spp.

Ash

Description Small or medium-size to large forest trees; leaves opposite, pinnately compound, the leaflets oblong or broadly elliptic to lanceolate, entire or serrate on the margins; flowers small, clustered or in short feathery racemes, the staminate and pistillate flowers sometimes on separate trees, appearing in spring; fruit a samara, with a single paddle-shaped wing.

WHITE ASH, *Fraxinus americana*

WHITE ASH, *Fraxinus americana*

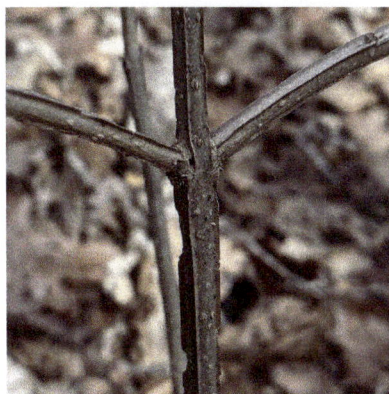

BLUE ASH, *Fraxinus quadrangulata*

Distribution Widely distributed. **Habitat** Glades, bluffs, and open woods, to low ground and floodplain forests.

The common **white** or **American ash**, *Fraxinus americana*, has smooth-edged ovate or oblong leaflets, somewhat whitish beneath, and U-shaped leaf scars. **Green ash**, *Fraxinus pennsylvanica*, is distinguished by glabrous twigs, lanceolate leaflets, which are serrate, and leaf scars not U-shaped but straight across the upper margin. **Blue ash**, *Fraxinus quadrangulata*, of dry bluffs and glades is easily identified by square twigs and pointed leaflets serrate on the margin.

Importance Ash fruits are eaten by wood ducks and are an important food of quail. Deer sometimes browse white ash heavily. Ash may be browsed intensively when cattle concentrate in bottoms.

Honeylocust

Description Medium-size to occasionally large tree with thorny trunk and branches; leaves alternate, pinnate, also twicepinnately compound, the numerous leaflets ovate-oblong, about ½ in. (10–12 mm.) long, sessile, with smooth margins; flowers small, greenish, in short drooping spikelike clusters, appearing in spring; pod large, thin, flattened, becoming twisted, to 12 in. (30 cm.) long or more, with numerous seeds.

HONEYLOCUST, *Gleditsia triacanthos*

Distribution Widely distributed.
Habitat Open woodland, pastures, bottom ground, and streambanks.
Importance The seeds and pulpy seed pods of honeylocust provide fall and winter food for deer and squirrels. Cattle will browse twigs of honeylocust in winter, and cattle and hogs eat the nutritious seeds in autumn.

Gymnocladus dioicus

Kentucky Coffeetree

Description Tall slender forest tree with somewhat pinkish-gray bark, stout twigs, and large leaf scars; leaves large, alternate, twice-pinnately compound, the numerous leaflets ovate, with pointed apex on short stalks, and smooth margins; flowers small, whitish, staminate and pistillate on separate trees, appearing in spring; pod heavy, thick, about 4½ in. (12 cm.) long and 1½ in. (4 cm.) wide, the several large seeds in a greenish gelatinous matrix.

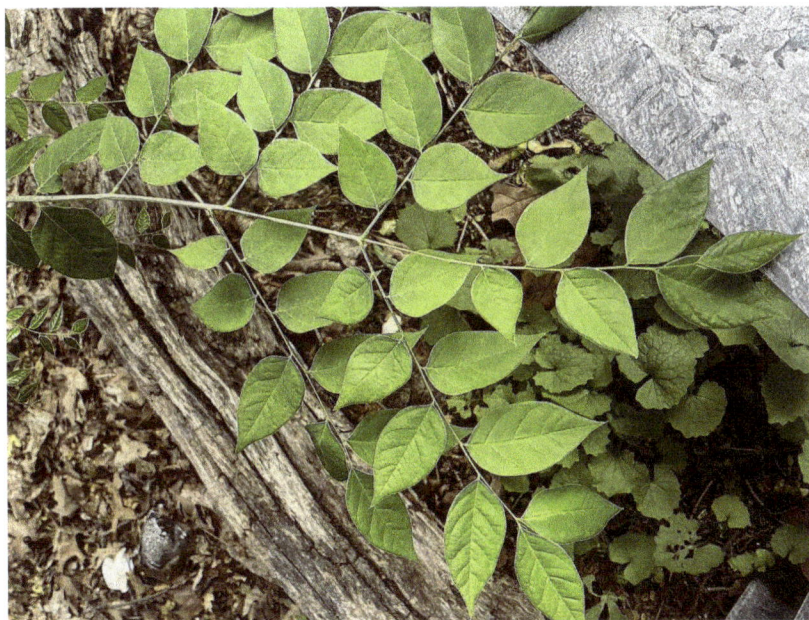

COFFEETREE, *Gymnocladus dioicus*

Distribution Widely scattered throughout the region, seldom abundant.

Habitat Ravines, lower slopes, and stream bottoms; on deep soils.

Importance The seeds and pulpy seed pods are eaten by deer and squirrels during autumn, winter, and into early spring. Kentucky coffeetree is seldom browsed by livestock, but cattle and hogs will eat the seeds. The greenish pulp of the unripe seed pods is thought to be poisonous to sheep, cattle, and horses, but few cases have been reported.

Ozark Witch-hazel

Description Shrubs to about 8–9 ft. (2½ m.) tall, producing new shoots from the base; twigs mostly pubescent; leaves simple, alternate, broadly elliptic with obtuse apex and rounded to tapering base, scalloped on the margin; flowers yellowish with reddish tinge, with 4 stringy petals, developing in winter and early spring; fruit a dry, woody capsule, 2-seeded.

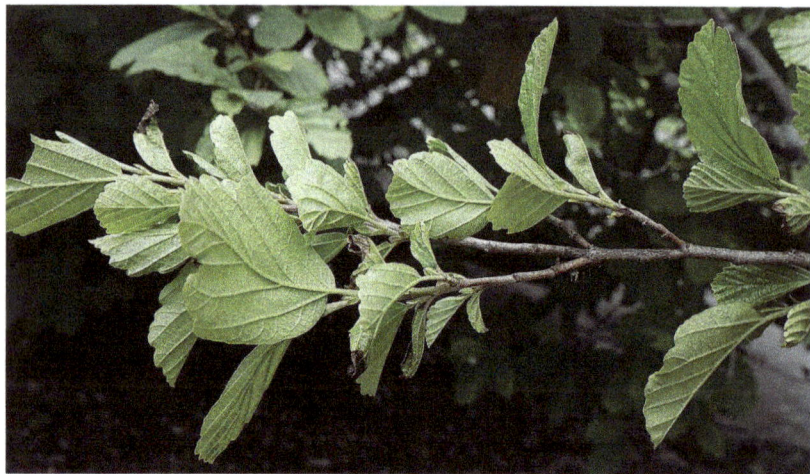

OZARK WITCH-HAZEL
Hamemelis vernalis LEFT, ABOVE

EASTERN WITCH-HAZEL
Hamemelis virginiana

Distribution Widely distributed.
Habitat Primarily rocky banks and along streams.

Eastern witch-hazel, *Hamamelis virginiana*, is a less common species, occurring only in the eastern Ozarks. It is easily separated from the Ozark witch-hazel by the twigs being smooth or nearly so, and by pale-yellow flowers, which appear in fall and winter. **Importance** Witch-hazel provides a small amount of winter food for deer and turkeys.

Hydrangea arborescens

Wild Hydrangea

Description Small to medium-sized shrub with branching stems, usually 3–6 ft. (1–2 m.) tall, and with loose scaly bark; leaves simple, opposite, ovate to cordate, long-petioled, serrate; flowers whitish, crowded in flat-topped clusters, the outer flowers showy but sterile, appearing in spring; fruit a dry capsule.

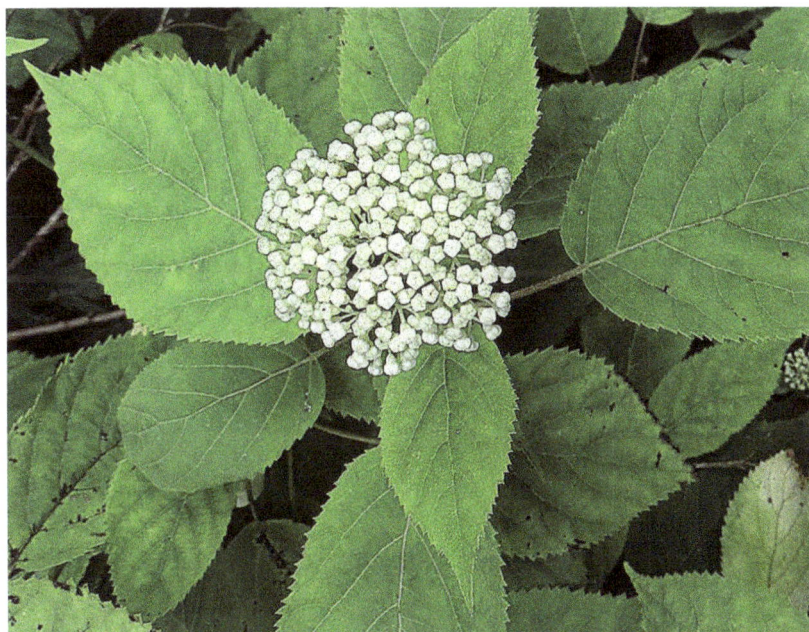

WILD HYDRANGEA
Hydrangea arborescens

Distribution Widely distributed.
Habitat Shaded bluffs and slopes; usually on limestone soils.
Importance Wild hydrangea is considered excellent deer food during the growing season. Since it is scattered and not generally abundant, it does not constitute a large part of the deer's diet. Under continued browsing it loses vigor and eventually dies and may be an early indicator of overbrowsing. Turkeys are re-

ported to pick hydrangea leaves in summer and fall. Wild hydrangea is poor forage for domestic livestock. It may cause poisoning in livestock, but few cases have been reported.

St. John's-wort

Description Upright annual and perennial herbs or shrubs, the latter to 6–7 ft. (2 m.) tall; leaves simple, opposite, linear, elliptic, or oblong, sessile, with entire margins, some herbaceous species with transparent (pellucid) dots on the leaves, visible when held to light; flowers bright yellow, with 5 petals and usually numerous stamens, appearing in summer; fruit a dry many-seeded capsule.

Hypericum perforatum

Hypericum punctatum

Distribution Widespread and sometimes common and abundant.
Habitat Open woods, fields, and waste areas, also streambanks and low ground; *Hypericum perforatum* is introduced from Europe.

Hypericum perforatum and the native *Hypericum punctatum* are common perennial herbs to 3 ft. (90 cm.) high, the first species identified by profuse branching, the second species sparingly branched or with a simple stem; leaves of both plants elliptic-oblong with pellucid dots, *H. punctatum* also with black dots.

Hypericum sphaerocarpum with linear leaves is a low herb about 20 in. (50 cm.) tall; it can be distinguished from the above species by its creeping rootstocks.

A common shrub of upland woods, as well as lower ground, is **shrubby St. John's-wort,** *Hypericum prolificum*, a coarse plant to 6–7 ft. (2 m.) tall or more with spatulate leaves, wedge-shaped toward the base of the blade.

Other species of various habitats also occur in the region, including two low annual species with distinctively narrow or scalelike leaves. These are nits-and-lice, *Hypericum drummondii*, and pine-weed, *Hypericum gentianoides*.

Hypericum spp.

St. John's-wort

Hypericum sphaerocarpum

SHRUBBY ST. JOHN'S-WORT
Hypericum prolificum

Importance Deer browse shrubby St. John's-wort during winter. Quail eat small amounts of nits-and-lice seed.

Mature woody plants are unpalatable to livestock. Young shoots are fair forage for goats and may occasionally be eaten by cattle and sheep. Generally they are taken only when no other forage is available. When eaten in large quantities, the plants are poisonous to cattle, sheep, and horses, particularly in late summer. The unpigmented skin of animals eating these plants may be photosensitized by sunlight. Blisters and scabby conditions develop on the face and ears and occasionally over the back and sides. Sick animals lose weight and in severe cases may become blind, develop sore mouths, and die from malnutrition. The poisonous effect is cumulative.

Mechanical control such as digging and mowing is usually practical only on small areas. Burning may cause the plants to spread. Judicious spraying with herbicides may be more feasible.

St. Andrew's Cross

Description Low leafy shrub with several dark slender stems clumped together, to about 15 in. (35 cm.) tall; leaves simple, opposite, linear-oblong, ⅜ in. (1 cm.) long or more, somewhat evergreen, with entire margins; flowers yellow, cross-shaped, with 4 petals, appearing in summer; fruit a small capsule with numerous seeds.

ST. ANDREW'S CROSS
Hypericum hypericoides

Synonym *Ascyrum hypericoides*
Distribution Widely distributed.
Habitat Oak and pine woods on dry slopes and ridges with cherty or sandy, generally acidic soils.
Importance The stems and leaves of *Hypericum* are often browsed by deer during the winter when other food is scarce. Thus, even though these plants are small and scattered and provide only a small amount of forage, they are an important component of deer range. Cattle will browse this plant when other forage is scarce.

Juglans spp.

Butternut, Walnut

Description Forest trees with heavy twigs and chambered pith; leaves large, alternate, pinnately compound, the leaflets more than 10, oblong-lanceolate, with pointed tip and rounded base, and with finely serrate margins; staminate flowers in drooping catkins, the pistillate flowers in small clusters at ends of branches; nut globose or oblong with leathery nonsplitting husk.

BLACK WALNUT, *Juglans nigra*

BUTTERNUT, *Juglans cinerea*

Distribution Widely distributed. **Habitat** Rich woods, ravines, streambanks, and bottomlands. The only two species are **black walnut**, *Juglans nigra*, with rounded nuts and light-brown pith, and **butternut**, *Juglans cinerea*, distinguished by oblong nuts and dark-brown pith. **Importance** Black walnut and, to a lesser extent, butternut, provide food for squirrels from the time the nuts ripen until supplies are exhauste. The nuts are often stored by squirrels for later use. Cattle will occasionally browse the leaves and twigs in late summer and fall when other feed is scarce. Hogs will readily eat the nuts.

Black walnut is an especially valuable tree for lumber, and the nuts of both species are often gathered for human consumption.

Ashe Juniper, Eastern Redcedar

Description Generally small or medium-sized evergreens with full crown and spreading branches; leaves numerous, overlapping, scalelike and sharp-pointed, about 1/16 in. (1½ mm.) long; fruits bluish, smooth, spherical, with 1 or 2 seeds.

EASTERN REDCEDAR
Juniperus virginiana

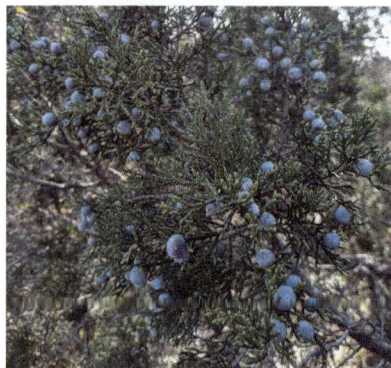

ASHE JUNIPER
Juniperus ashei

Distribution Widespread, one species with limited range.
Habitat Old fields, pastures, glades; on dry thin soils and limestone outcrops.

The more restricted **Ashe juniper**, *Juniperus ashei*, occurs locally in the western Ozarks. It is distinguished from the familiar and widespread **eastern redcedar**, *Juniperus virginiana*, by its shorter, somewhat bushy habit, occasionally with several trunks from ground level, and its larger, sharply pointed unpitted seeds.

Importance Redcedar is browsed intensively during winter and spring when deer populations are high.

The fruits make up a small part of the diet of ruffed grouse, quail, prairie chickens, pheasants, and probably many nongame birds. The thick crowns provide nesting and roosting cover for many birds, and dense thickets make good escape cover for deer.

Lindera benzoin

Spicebush

Description Shrub to about 15 ft. (5 m.) tall; twigs and bark with spicy or aromatic flavor; leaves simple, alternate, narrow-ovate with wedge-shaped base and pointed tip, dark green above, pale beneath, entire or nearly so on the margins; flowers small, yellow, in clusters, appearing before the leaves in early spring; fruit a drupe, bright-red.

SPICEBUSH, *Lindera benzoin*

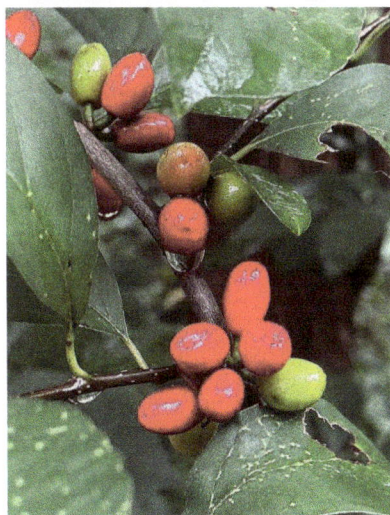

Distribution Widely distributed. **Habitat** Low or damp woods, ravines, and small spring-fed streams.

Importance Spicebush provides a small amount of winter food for deer.

Sweetgum, Redgum

Description Large forest tree with gray bark and sticky sap, the twigs and branches often with corky wings or ridges; leaves simple, alternate, fragrant, about as broad as long with 5 long-pointed lobes minutely toothed; flowers small, greenish, the pistillate ones in ball-shaped clusters on a long stalk, in spring; fruit spherical.

SWEETGUM, *Liquidambar styraciflua*

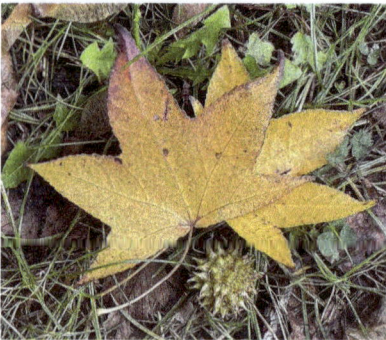

Distribution Eastern and southern part of the Ozarks.
Habitat Flat woods, swampy ground, and along streams.
Importance Sweetgum buds and seeds provide auxiliary food for squirrels. Cattle concentrated in bottom areas will browse the twigs to a limited extent.

Honeysuckle

Description Trailing or climbing vines, or somewhat bushy or shrublike; leaves simple, opposite, rounded or ovate, entire on margins; flowers showy, tubular, yellow white to yellow or orange yellow, appearing in midspring and summer; fruit small, spherical-oblong, several-seeded.

JAPANESE HONEYSUCKLE
Lonicera japonica TOP, ABOVE

YELLOW HONEYSUCKLE
Lonicera flava

Distribution Widely scattered throughout the region, occasionally abundant.

Habitat Open woods, roadsides, and waste areas.

Yellow honeysuckle, *Lonicera flava*, is a widespread native plant, somewhat shrubby, with thick sessile leaves, those subtending the flowers joined at the base and surrounding the stem.

Japanese honeysuckle, *Lonicera japonica*, is an escape from cultivation, and differs by its sprawling or climbing vinelike habit, petioled leaves, and pubescent upper stems.

Importance Honeysuckle is occasionally browsed by deer in the Ozarks.

MORACEAE, MULBERRY FAMILY

Osage-orange, Bois-d'arc, Hedge-apple

Description Medium-sized tree with spiny branches, milky sap, and bright orange inner bark; leaves simple, alternate, ovate to lanceolate with pointed apex, shiny, dark green, entire on the margins; flowers small, in dense clusters, developing in spring; fruit large, spherical, greenish, about 4 in. (10 cm.) in diameter, with numerous seeds.

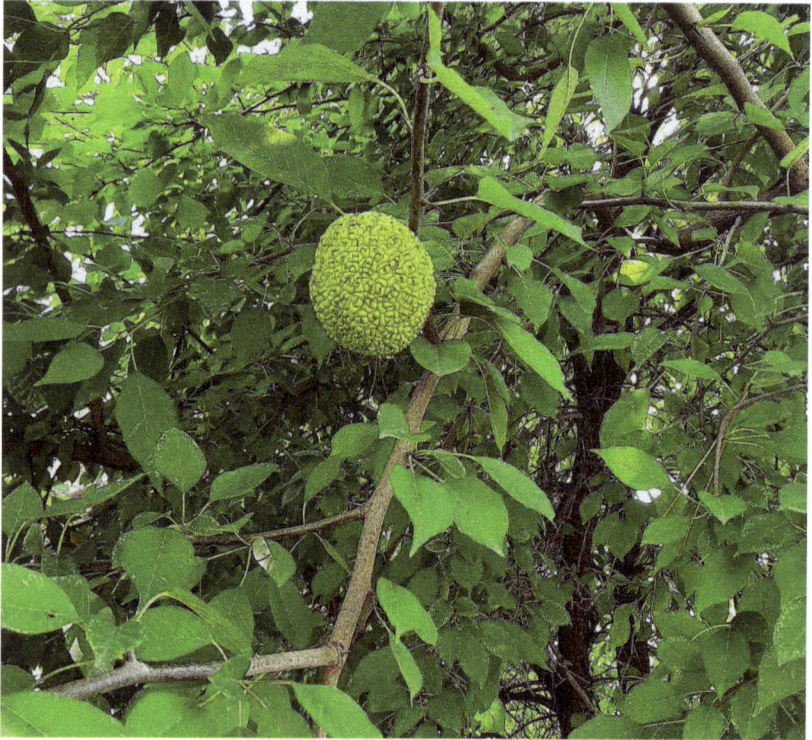

OSAGE-ORANGE, *Maclura pomifera*

Distribution Widespread throughout the range, but probably introduced in the central and northern sections.

Habitat Low woods, pastures, fence rows, frequently spreading to open ground.

Importance The fruits of osage-orange are readily eaten by fox squirrels. The seeds make up a small part of the diet of quail, and raccoons obtain some food from this plant. Osage-orange twigs are occasionally browsed by cattle in areas where forage is scarce.

Osage-orange has excellent properties as fenceposts for which it is still used.

Some people develop a dermatitis when in contact with the milky sap from the stems, leaves, and fruits.

Malus ioensis

Crabapple, Wild Crab

Description Small tree with rounded crown and spreading branches; leaves simple, alternate, ovate-elliptic, coarsely toothed, the vertical side of the lobes more or less parallel to the midrib; flowers large, pink, fragrant, with 5 petals, in midspring; apple fruit about 1 in. (2½ cm.) in diameter or less, somewhat aromatic.

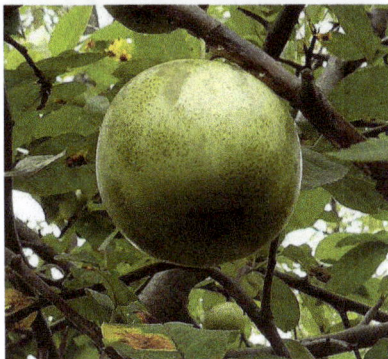

CRABAPPLE, *Malus ioensis*

Synonym *Pyrus ioensis*
Distribution Widely distributed.
Habitat Prairies, woods openings, thickets, pastures, bottomlands.
Importance Deer eat crabapple

fruits in summer and autumn. Cattle will browse the leaves and twigs to some extent on areas where forage is scarce. Hogs relish the fruits in autumn.

Moonseed

Description Slender twining vine, lacking tendrils, with glabrous stems; leaves simple, alternate, with 3–7 shallow lobes, the long petiole attached to the underside of the blade near the lower edge; flowers small, greenish white, appearing in spring and summer; fruits blackish, grapelike, with a single crescent-shaped bony seed.

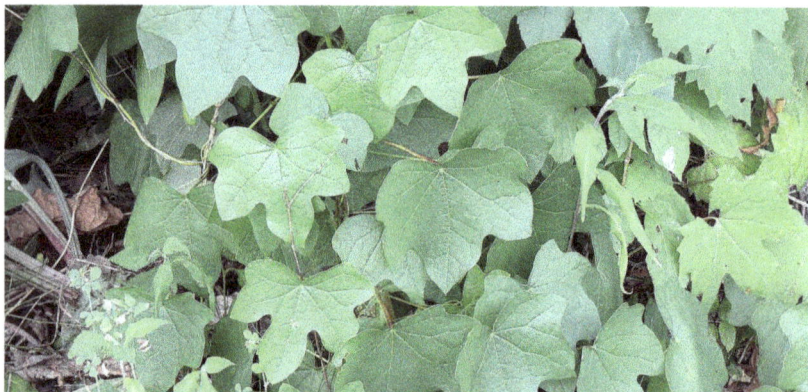

MOONSEED, *Menispermum canadense*
ABOVE, LEFT

Distribution Widely distributed but usually not abundant.

Habitat Low woods, bottomlands, streambanks, and rich ground.

Carolina moonseed, *Cocculus carolinus,* is a slender climbing vine somewhat similar in general appearance to moonseed but with ovate or heart-shaped leaves and scarlet fruits.

Importance Moonseed fruits may be mistaken for wild grapes and, if eaten by humans, can cause mild poisoning. The sharp ridges of the fruit pits may also cause mechanical injury to the intestinal tract. Moonseed is seldom eaten by livestock or wildlife.

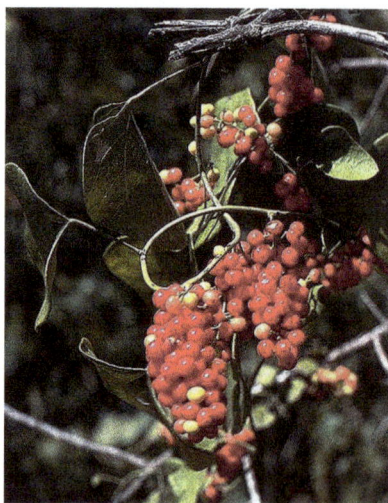

CAROLINA MOONSEED
Cocculus carolinus

Morus rubra

Red Mulberry

Description Occasionally large trees; leaves simple, alternate, broadly ovate to irregularly lobed with short more or less blunt teeth on the margins; flowers small, staminate and pistillate on separate trees, appearing in spring; fruit berrylike, somewhat oblong, purplish-black at maturity, about 1 in. (2½ cm.) long.

WHITE MULBERRY, *Morus alba*

RED MULBERRY, *Morus rubra*
TOP, ABOVE

Distribution Widely distributed. The introduced **white mulberry, *Morus alba*** is also present and its leaves are widely used in Asia as food for silkworms.
Habitat Upland woods, pastures, waste ground, and bottomlands.
Importance Where available, red mulberry fruits make up a considerable part of the diet of squirrels from May through July. Raccoons, wild turkeys, and ruffed grouse also eat varying amounts of the fruits, as do many nongame birds. Red mulberry is browsed to some extent by cattle in the fall.

Blackgum, Black Tupelo

Description Medium-sized to large tree with spreading branches; leaves simple, alternate, obovate, tapering to short petiole, shiny green above, turning scarlet in autumn, entire on the margins; flowers small, the staminate numerous, the pistillate few, in clusters, developing in spring; fruit dark blue, fleshy, about ½ in. (10–12 mm.) long, on a long stalk.

BLACKGUM, *Nyssa sylvatica*

Distribution Widely distributed.
Habitat Cherty or sandy slopes and ridges; on acid soils.
Importance Deer browse blackgum leaves and twigs during summer, and eat the fruits during autumn. Squirrels and turkeys eat the fruits during September and October. Blackgum fruits are considered a preferred turkey food.

Ironwood, Eastern Hophornbeam

Description Small tree with thin flaky bark, smooth slender twigs, and shiny buds; leaves simple, alternate, oblong-oblanceolate, cordate or rounded at base, pointed at apex, sharply serrate on the margin; staminate flowers in drooping catkins, the pistillate flowers in clusters at ends of branches; fruit a smooth nutlet in a flat bladder, in hoplike clusters.

IRONWOOD, *Ostrya virginiana*

Distribution Widely distributed.
Habitat Open woods, on dry or rocky soils, also moist woods or along streams.
Importance Catkins and buds of ironwood rank as the most important grouse food by volume consumed during late autumn, winter, and early spring. Quail occasionally eat the seeds, and they are a preferred winter food of turkeys.

Virginia Creeper, Woodbine

Description Woody climber with tendrils attaching by minute suction-like disks, also spreading on ground and forming dense cover; leaves alternate, palmately compound, long-petioled, leaflets mostly 5, elliptic to obovate, coarsely serrate on the margin; flowers small, greenish, numerous, appearing in late spring and summer; fruit a small, blackish berry about 3/16 in. (4–5 mm.) in diameter, with 1–4 seeds.

VIRGINIA CREEPER, *Parthenocissus quinquefolia*

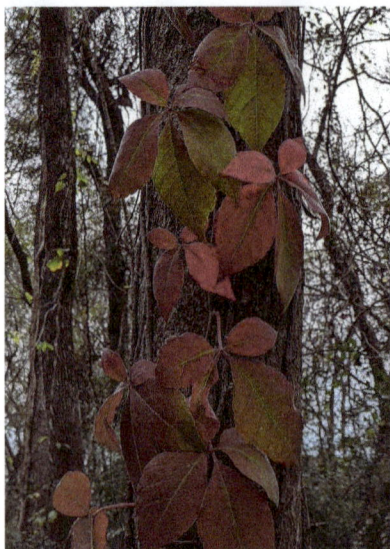

Distribution Widespread and sometimes common and abundant.
Habitat Woodland sites, occasionally abundant in local habitats,
 Poison ivy (*Toxicodendron* spp.) with which the species may sometimes be confused has 3 leaflets and small yellowish fruits, and lacks the suction disks of Virginia creeper.
Importance Deer browse the leaves and stems of woodbine during spring and summer, and eat the fruits during autumn. Quail consume minor amounts of the fruit. Squirrels eat the bark in winter and leaves and fruit in summer. Many species of birds eat the fruits.

Pinus echinata

Shortleaf Pine

Description Tall forest species of the hard pine group characterized by largeplated reddish bark on mature trees; young twigs whitish pink; needles to 3–4 in. (8–10 cm.) long, 2 or 3 together in a narrow bundle; cones about 3 in. (8 cm.) long; cone scales with a slight prickle.

SHORTLEAF PINE, *Pinus echinata*

Distribution Wide-ranging to the south, restricted at northern limits in Missouri Ozarks.

Habitat Mostly upland slopes and dry ridges; on sandy, cherty, or igneous soils and outcrops; the only native pine of the Ozark region.

Importance Scattered thickets of young pine furnish valuable wildlife cover. Occasionally pines are browsed by deer when more desirable foods are scarce. Young trees, transplanted from nursery beds, are especially susceptible to deer browsing damage during winter. Turkeys occasionally eat pine seeds.

Cattle, if confined to an area where forage is scarce, may browse pine in late winter.

Cherry, Plum

Description Shrubs with sharp or thorny branches, sometimes forming dense thickets, or small to medium-sized trzes, mostly solitary; leaves simple, alternate; elliptic, obovate, or oblong-lanceolate, frequently with glandular petioles, finely serrate; flowers white, with 5 petals, in showy clusters or racemes; fruit fleshy, with a single stonelike seed.

BLACK CHERRY, *Prunus serotina*

Distribution Widespread, one species with limited range.
Habitat Prairies, open woodland, pastures, waste ground, bottomlands, and streambanks.

Black cherry, *Prunus serotina*, is identified by leathery lanceolate leaves and small incurved teeth on the margin, flowers in drooping racemes and small dark-colored fruits. The **chokecherry**, *Prunus virginiana*, with thin obovate leaves, occurs primarily at the northern fringes of the Ozark region. The flowers of these two species appear after the leaves have developed.

Several species of **wild plum** are common, including *Prunus americana* and *Prunus hortulana*, both producing red fruits when ripe, the first species frequently forming dense thickets. These species generally have "spurlike" branches, and the flowers appear before the leaves.

Importance Black cherry fruits are eaten by wild turkeys, squirrels, raccoons, quail, and ruffed grouse. In the Ozarks, black cherry is only occasionally browsed by deer. Fruits of wild plum furnish summer food for squirrels, deer, raccoons, and quail. The less common chokecherry is occasionally eaten by squirrels

Prunus spp.

Cherry, Plum

CHOKECHERRY, *Prunus virginiana*

WILD PLUM, *Prunus americana*

CHOKECHERRY, *Prunus virginiana*

during summer. Also, many species of nongame birds eat the fruits of cherry and plum.

Black cherry and chokecherry are only moderately palatable for cattle and sheep and may be poisonous. The toxic substance, prussic acid, is present in both leaves and the fruit stones. Several cases have been reported of humans being poisoned from eating the fruit stones, while many cases of livestock poisoning have resulted from browsing the foliage. The foliage is dangerous over a long period of time although the plants are usually harmless by October. The leaves of the very young shoots are the most palatable and also the most harmful. Freshly wilted leaves may be especially toxic.

Oak

Description A large, complex genus in the temperate zone of the world; in our region varying from small scrub types on marginal sites, to tall forest trees on deep soils. Leaves simple, alternate, lobed or bristly tipped, or with entire margins. Fruit an acorn.

Distribution Widespread, or some species with limited range.

Habitat Upland sites, ravines. bottomlands, and alluvium.

Common or widely distributed species of the **white oak group** are characterized by deeply or shallowly lobed leaves without bristly tips and by acorns maturing in one season. This group includes **white oak**, *Quercus alba*, with usually narrow elongate lobes; **bur oak**, *Quercus macrocarpa*, with obovate leaves, shallowly lobed or merely sinuous, the deepest sinuses at the middle of the blade; **post oak**, *Quercus stellata*, with wide more or less cross-shaped lobing, and **chinkapin oak**, *Quercus muehlenbergii*, with coarsely dentate, oblong-lanceolate leaves, the lateral veins more or less parallel from the midrib to the margin.

In the **black oak group**, the species are characterized by bristly tipped leaves and angular lobes, and require two seasons for development of the acorn. This group includes **shingle oak**, *Quercus imbricaria*, with elliptic-oblong, unlobed leaves, more prevalent in the northern parts of the range; **blackjack oak**, *Quercus marilandica*, the leaves leathery, coarse-textured,

broadly obovate, more or less shallowly 3-lobed and tapering to base; **northern red oak**, *Quercus rubra*, the leaves with 7 or more spiny lobes, each about ¼ as long as the leaf is wide, mostly glabrous beneath; **black oak**, *Quercus velutina*, with leaves about 7-lobed, brownish pubescent or somewhat scurfy beneath, and with yellow inner bark; **Shumard oak**, *Quercus shumardii*, with deep somewhat rounded sinuses and 7 to 9 lobes, the entire leaf about as wide as long.

Other less common oaks include **scarlet oak**, *Quercus coccinea*, of the eastern Ozarks, **southern red oak** or **Spanish oak**, *Quercus falcata*, of the southern part of the range, and **pin oak**, *Quercus palustris*, of low ground and alluvium, generally not common in the Ozark range. Other oak species are present in the Ozark region as well.

Importance When available, acorns are the single most important wildlife food in the Ozarks. Good wildlife reproduction usually follows years of acorn abundance, and poor reproduction follows years of scarcity among species dependent upon acorns for food.

Acorns are especially important for wild turkeys and are the single

Oak

WHITE OAK, *Quercus alba*
WHITE OAK GROUP

BUR OAK, *Quercus macrocarpa*
WHITE OAK GROUP

most important food during winter. Squirrels eat acorns when available; they also eat oak buds during spring.

Acorns are considered the most important food for deer. Oak leaves and twigs are occasionally eaten, especially those of white oak. Ruffed grouse eat acorns throughout the year. Wood ducks, quail, and raccoons also eat a considerable amount of acorns.

Acorns are also eaten by many other animals, notably blue-jays and redheaded woodpeckers, which may take a considerable part of the annual production before it falls from the trees.

The opening buds and young leaves of several species of oak supply limited amounts of spring forage to cattle, sheep, and goats. The forage is generally rated poor to fair for most classes of livestock, although the nutrient content is good. Cattle also eat the leaves and twigs in summer when salt is scarce. Goats browse the young sprouts throughout the year.

Non-fatal stock poisoning has been reported from eating the leaves, usually in the spring when other forage is scarce. The leaves are not injurious, however, when eaten in mixture with other forage. Apparently goats are not injured when they eat the leaves and twigs.

Hogs relish the acorns, which have a very high fat and oil content, but a low protein content.

Oak

POST OAK, *Quercus stellata*
WHITE OAK GROUP

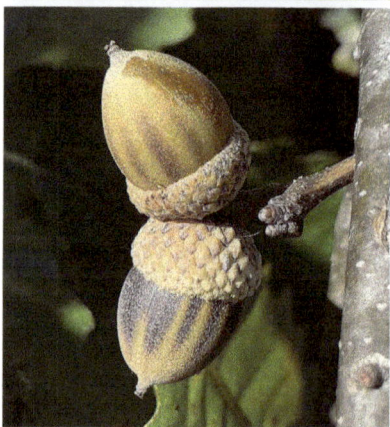

CHINKAPIN OAK
Quercus muehlenbergii
WHITE OAK GROUP

SHINGLE OAK
Quercus imbricaria
BLACK OAK GROUP
LEFT, ABOVE

Quercus spp.
Oak

BLACKJACK OAK
Quercus marilandica
BLACK OAK GROUP
LEFT, ABOVE

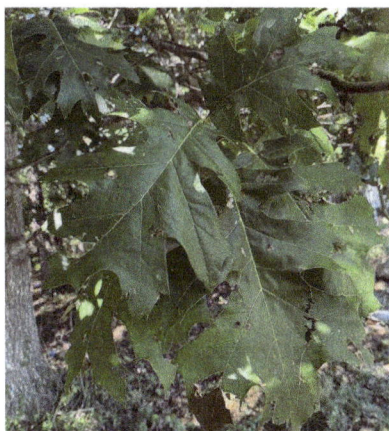

BLACK OAK
Quercus velutina
BLACK OAK GROUP

NORTHERN RED OAK
Quercus rubra
BLACK OAK GROUP

SHUMARD OAK
Quercus shumardii
BLACK OAK GROUP
RIGHT

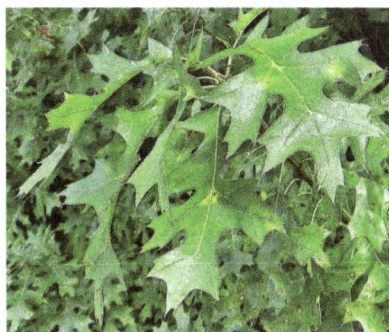

Buckthorn, Indian-cherry

Description Shrubs or small trees; leaves simple, alternate, elliptic to oblong-lanceolate, appearing entire or with minute teeth on the margins; flowers minute, greenish, in small axillary clusters, appearing in spring; fruit black, berrylike; 2- or 3-seeded.

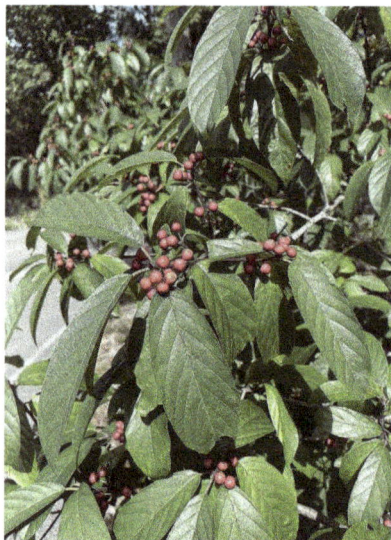

INDIAN CHERRY
Rhamnus caroliniana

Distribution Widely distributed.
Habitat Glades, thickets, and open woods on mainly calcareous soils, or in stream bottoms.

Two common species are **lance-leaved buckthorn**, *Rhamnus lanceolata*, a small shrub, and **Indian cherry** or **Carolina buckthorn**, *Rhamnus caroliniana*, larger and treelike, with brownish hairy buds lacking scales.

Importance Deer and cattle occasionally browse Carolina buckthorn during winter. Studies in other regions show that the fruits are eaten by many species of birds, especially the pileated woodpecker.

LANCE-LEAVED BUCKTHORN
Rhamnus lanceolata
TOP, ABOVE

Rhus spp., *Toxicodendron* spp.

Sumac, Poison Ivy

Description Erect and shrubby, or climbing vinelike plants; leaves alternate, trifoliate, or pinnately compound with numerous ovate to lanceolate leaflets; flowers small, greenish yellow, clustered, appearing before or with the leaves; fruits small, reddish to cream colored, 1-seeded, in more or less dense clusters.

POISON IVY, *Toxicodendron radicans*

Distribution Widespread and sometimes common and abundant.
Habitat In various habitats ranging from upland woods, glades, and fields to low woods and streambanks.

Poison ivy, *Toxicodendron radicans*, is identified by trifoliate leaves and smooth yellowish berries, and varies in habit from low to mediumsized shrubs to high-climbing vines with aerial roots. The shrubby **fragrant sumac**, *Rhus aromatica*, also has 3 leaflets, but these are fragrant, nontoxic, and all sessile; the fruits are reddish, pubescent, and berrylike.

Two common sumacs of thickets, prairies, and glades, with pinnately compound leaves, are **smooth sumac**, *Rhus glabra*, with serrate leaflets turning crimson in the fall; and **winged sumac**, *Rhus copallina*, with leaflets smooth-edged and the rachis conspicuously winged, also turning red in autumn. All poisonous species have white or yellowish fruits, whereas nontoxic species have reddish fruits.

Sumac, Poison Ivy

FRAGRANT SUMAC, *Rhus aromatica*

Importance Sumacs rank as the fourth most important deer food in Missouri by volume consumed. Winged and smooth sumac twigs and fruits are browsed throughout the year, but the greatest volume is taken during autumn and winter. Fragrant sumac twigs and fruit are browsed during spring and summer. Deer occasionally browse poison ivy during winter.

Ruffed grouse eat a considerable volume of fragrant sumac catkins during winter and spring, and fruits during May and June. Smooth sumac and poison ivy fruits are used to a lesser extent but are still rated as important grouse foods. Winged sumac provides food for prairie chickens, primarily during winter but also to a lesser extent during spring and autumn. Quail eat the fruits of winged sumac, smooth sumac, and poison ivy during the critical late winter period. Sumacs and poison ivy also make up a small part of the diet of turkeys, pheasants, wood ducks, and squirrels. Field observations show that sumac bark often provides sustaining food for rabbits

Rhus spp., *Toxicodendron* spp.

Sumac, Poison Ivy

SMOOTH SUMAC, *Rhus glabra*

WINGED SUMAC, *Rhus copallina*

during severe winter weather with heavy snow.

The sumacs vary from poor to good forage for cattle and sheep but are good for goats. Cattle will graze the twigs in winter and the new leaves and flowers in the spring.

Poison ivy contains a toxic sub-stance which may cause a severe in-flammation and blistering to the skin of humans. The degree of poi-soning varies, depending on the condition of the plant, the degree of exposure, and the susceptibility of the person.

Currant, Gooseberry

Description Spiny understory shrubs; leaves simple, alternate, lobed, mostly less than 2 in. (5 cm.) wide, with serrate or scalloped edges; flowers small, whitish, pendulous, from slender stalks, appearing in spring; fruits spherical, purplish black at maturity, smooth or prickly.

PRICKLY GOOSEBERRY
Ribes cynosbati

Distribution Widely scattered throughout the region, seldom abundant.

Habitat Woodland habitats, pastures, and low ground; on mostly neutral or alkaline soils.

The common **Missouri gooseberry**, *Ribes missouriense*, has smooth fruits, differing from the less common **prickly gooseberry**, *Ribes cynosbati*, with spiny fruits.

A less common species with conspicuous yellow flowers and lacking spines on the stem is the **golden currant**, *Ribes aureum*, primarily of the western Ozark region, but occasionally cultivated elsewhere.

Importance The currants are considered a poor to fair forage for cattle and fair to good forage for sheep and goats. The fruits and leaves are eaten occasionally by squirrels during summer.

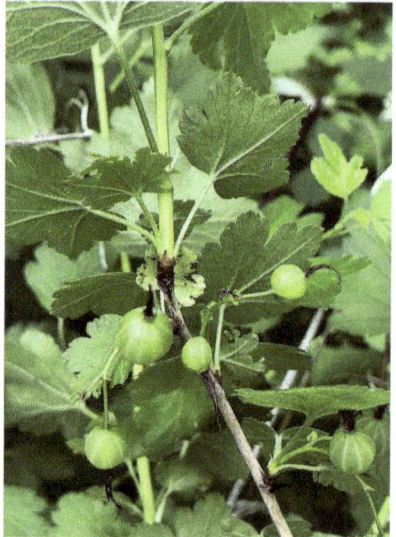

MISSOURI GOOSEBERRY
Ribes missouriense
TOP, ABOVE

The currants were cultivated for their fruits in the past but little grown today.

Ribes spp.

Currant, Gooseberry

GOLDEN CURRANT, *Ribes aureum*

GOLDEN CURRANT, *Ribes aureum*

Black Locust

Description Small to medium-sized tree of the legume family; leaves alternate, pinnately compound, the leaflets oblong, entire on the margins; short spinelike stipules at the base of leaf stalk; flowers showy, white, scented, forming elongate racemes, appearing in midspring; pods mostly flat, about 4–5 in. (10–12 cm.) long, with several seeds.

BLACK LOCUST
Robinia pseudoacacia

Distribution Widely scattered throughout the region, occasionally abundant.

Habitat On neutral or alkaline soils; extensively used in cover and erosion control plantings.

Importance In a study of overpopulated ranges in northern Arkansas, deer ate over 80 percent of the season's growth of leaves and twigs of black locust. Little deer use has is observed on high-quality Ozark ranges. Seeds of this plant are occasionally eaten by squirrels and quail.

Black locust is considered worthless to poor as a forage plant for domestic livestock. The bark or young shoots may fatally poison horses, mules, cattle, and sheep.

Rose

Description Plants mostly prickly, low, and bushy, or tall, somewhat climbing or trailing with long, arching canes; leaves alternate, compound, stipulate, with 3 to 11 ovate to lanceolate or oblong leaflets, serrate on the margin; flowers pink or white with 5 petals, appearing in spring and summer; fruit red, subglobose, about 2/5 in. (1 cm.) long with several seedlike achenes.

CAROLINA ROSE, *Rosa carolina*

Distribution Widespread and sometimes common and abundant.
Habitat Open woods, glades, prairies, pastures, and waste areas. Multiflora rose, introduced from Asia as a fence-row planting and for erosion control, has spread and become a nuisance in many places.

The common **Carolina rose** or **pasture rose**, *Rosa carolina*, is a low shrub usually not exceeding 3 feet (90 cm.), with 5–7 leaflets and pink flowers.

A common **climbing rose** is *Rosa setigera*, with long trailing canes and rank habit, frequently with only 3 leaflets.

The **multiflora rose**, *Rosa multiflora*, naturalizes in fields and pastures from fence row plantings. It is distinguished by tall arching canes, 7–9 leaflets, and numerous white flowers in late spring and early summer.

Rosa spp.

Rose

CLIMBING ROSE, *Rosa setigera*

MULTIFLORA ROSE, *Rosa multiflora*
ABOVE, RIGHT

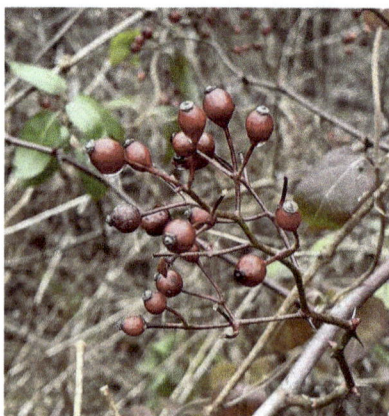

Importance Deer browse the twigs and fruits of wild rose throughout the year. Native roses and multi-flora rose provide year-round food for prairie chickens and ruffed grouse; the persistent fruits of these plants are also a readily available food during winter snows. Rose fruits also make up a small part of quail, pheasant, and turkey diets. Field observations show that wild rose plants are often eaten by rabbits. Roses are considered poor to fair forage for cattle, but fairly good for sheep and goats.

The roses often develop dense thickets, which make excellent cover for small birds and mammals.

Rubus spp.

Bramble

Description Plants thorny, shrubby, with stout upright or arching canes, or vinelike with trailing stems; leaves alternate, with 3–5 coarsely serrate leaflets; flowers white (for all species included here) with 5 petals, developing in middle or late spring; fruit a fleshy berry.

DEWBERRY, *Rubus fllagellaris*

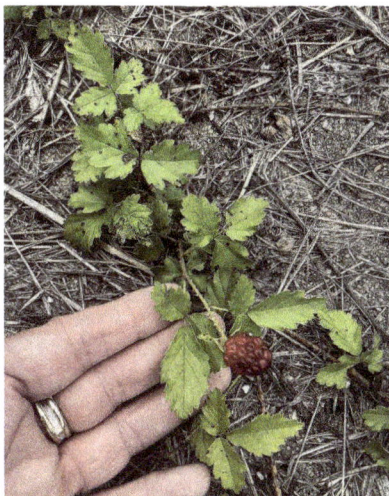

SOUTHERN DEWBERRY
Rubus trivialis

Distribution Widespread and sometimes common and abundant.
Habitat Open woods, glades, old fields, and waste areas to low moist ground or stream bottoms.

Dewberry, *Rubus flagellaris*, a most common species of old fields and open woods, is characterized by trailing habit and glabrous stems and leaf stalks. A less common species, also trailing, is the **southern dewberry**, *Rubus trivialis*, occurring mostly on lower ground, and distinguished by reddish bristly hairs on the stems and leaf stalks.

Two common shrubby types with stout upright canes are **high-bush blackberry**, *Rubus pensilvanicus*, and **black raspberry**, *Rubus occidentalis*. The leaflets of raspberries are whitish beneath, those of blackberries and dewberries are green or grayish green. Raspberries also are distinguished by the hollow thimblelike fruit whereas the fruit of blackberries and dewberries is solid.

Rubus spp.   ROSACEAE, ROSE FAMILY

Bramble

HIGH-BUSH BLACKBERRY
Rubus pensilvanicus
ABOVE, RIGHT

BLACK RASPBERRY, *Rubus occidentalis*

Importance Brambles provide food and valuable cover for many wildlife species. Deer eat the fruit and browse the tender primocanes during spring and summer. Much of the summer diet of wild turkeys is composed of these fruits. Prairie chickens and to a lesser extent quail, ruffed grouse, raccoons, pheasants, and squirrels obtain food from these plants. Rabbits and many nongame species eat the fruit, as indicated by records from other regions.

The blackberries are considered poor to fair forage for cattle and sheep. In addition to providing fruit that is highly relished by humans, brambles have some value in erosion control, forming dense thickets on barren and infertile ground.

Elderberry, American Elder

Description Tall shrub with soft woody stems and large white pith, the bark somewhat warty; leaves opposite, pinnately compound with lanceolate toothed leaflets; flowers small, white, fragrant, in large flat-topped clusters, appearing in spring and summer; fruits berrylike, purplish black, with 3 nutlets.

ELDERBERRY, *Sambucus nigra*

Synonymyn *Sambucus canadensis*
Distribution Widespread and generally common.
Habitat Open woods, thickets, streambanks and waste ground.
Importance Deer eat elderberry fruits during autumn. The fruit also provides occasional food for racoons, squirrels, turkeys, quail, pheasants, and many species of nongame birds.

Livestock seldom browse elderberry; however, cattle and sheep have been poisoned by eating new shoots, leaves, and opening buds.

The berries are often gathered to make wines, jellies, and pies.

Sassafras

Description Small to medium-sized tree forming thickets from suckers, or sometimes large solitary forest trees in favorable sites; twigs greenish, smooth, aromatic; leaves simple, alternate, of various shapes from ovate to three-lobed, mostly glabrous, with entire margins; flowers small, yellowish green, staminate and pistillate on separate trees, appearing in spring; fruits blue, ovoid, on thick red pedicels or stalks.

SASSAFRAS, *Sassafras albidum*

Distribution Widespread and sometimes common and abundant.

Habitat On various sites including old fields, pastures, wooded slopes and cove habitats.

Importance Sassafras is an important wildlife plant, partly because it is so abundant and widely distributed throughout the Ozarks. Deer browse the small twigs and large buds during winter, spring, and summer. Sassafras fruits are important quail food. Squirrels eat the fruits during August and September. Ruffed grouse and turkeys make occasional use of this species.

Sassafras has very little forage value for domestic livestock, being eaten only when better forage is not available or on winter range.

Traditionally, the roots were used to make sassafras tea, and the fragrant leaves were dried and powdered and added to gumbo soup for flavor. However, there is now some question as to the safety of ingesting sassafras.

Sideroxylon lanuginosum

Gum Bumelia

Description Usually a small tree, sometimes with thorny branches; leaves simple, alternate, with broad rounded apex and tapering base, glossy green above, rusty pubescent beneath, entire on the margins; flowers small, whitish, in clusters from axils of leaves, appearing in summer; fruit globular, dark, berrylike, about ¼ in. (6–7 mm.) across.

GUM BUMELIA
Sideroxylon lanuginosum

Synonym *Bumelia lanuginosa*
Distribution Widely distributed.
Habitat Open woods and glades; on dry rocky soils.
Importance Deer occasionally eat the fruit and leaves of this plant. Quail eat small amounts of the fruit during autumn. Cattle will occasionally browse the twigs on winter ranges.

Smilax spp.   SMILACACEAE, GREENBRIER FAMILY

Carrion-flower, Catbrier, Greenbrier

Description Herbaceous or woody vines, sometimes shrubby, generally with coiling tendrils and smooth or prickly stems; leaves simple, alternate, ovate or rounded to hastate or triangular, with conspicuous venation and mostly entire margins; flowers small; greenish or yellowish, in globose umbels, appearing in middle to late spring; fruit a small, bluish or black, 1 or 2 seeded berry.

BRISTLY GREENBRIER
Smilax tamnoides

Distribution Widely distributed.
Habitat Glades, open woods, ravines, low ground, and stream bottoms.

The common **bristly greenbrier**, *Smilax tamnoides* is identified by woody spiny stems and thin-textured ovate leaves, of low woods or moist slopes. **Saw greenbrier**, *Smilax bona-nox*, is also woody and spiny, but has leathery more or less triangular-elongate leaves, is commonly in rocky woods or glades but may also be on low ground.

Two common carrion-flowers with herbaceous stems and odoriferous flowers but lacking spines,

SAW GREENBRIER
Smilax bona-nox
top, above

and with more or less ovate to rounded leaves are **Blue Ridge carrion-flower**, *Smilax lasioneura*, a climbing vine; and **upright carrion-flower**, *Smilax ecirrhata*, a low brushy plant not exceeding 3.3 ft. (1 m.), with few or no tendrils.

Smilax spp.

Carrion-flower, Catbrier, Greenbrier

BLUE RIDGE CARRION-FLOWER
Smilax lasioneura
LEFT, ABOVE

UPRIGHT CARRION-FLOWER, *Smilax ecirrhata*

Importance Greenbrier is an important deer food and is often browsed heavily during spring and summer and usually to a lesser degree during winter. Leaves, stems and the persistent fruits are eaten. Greenbrier is a preferred food of turkeys.

Raccoons, pheasants, and ruffed grouse also get a small amount of food from these plants. Rabbits find good cover in old-field greenbrier thickets and also nibble the stems.

These plants have a low grazing value for domestic livestock.

Symphoricarpos orbiculatus **CAPRIFOLIACEAE, HONEYSUCKLE FAMILY**

Coralberry, Buckbrush, Indian Currant

Description Leafy shrubs to 3 ft. (90 cm.) tall, forming dense thickets; bark reddish brown, in loose peeling strips; leaves simple, opposite, oval to broadly elliptic, sessile or nearly so, entire-margined; flowers small, greenish, forming dense clusters in axils of leaves, appearing in summer; fruit a reddish berry about ¼ in. (5–6 mm.) long, 2-seeded, persistent in axillary clusters through the winter.

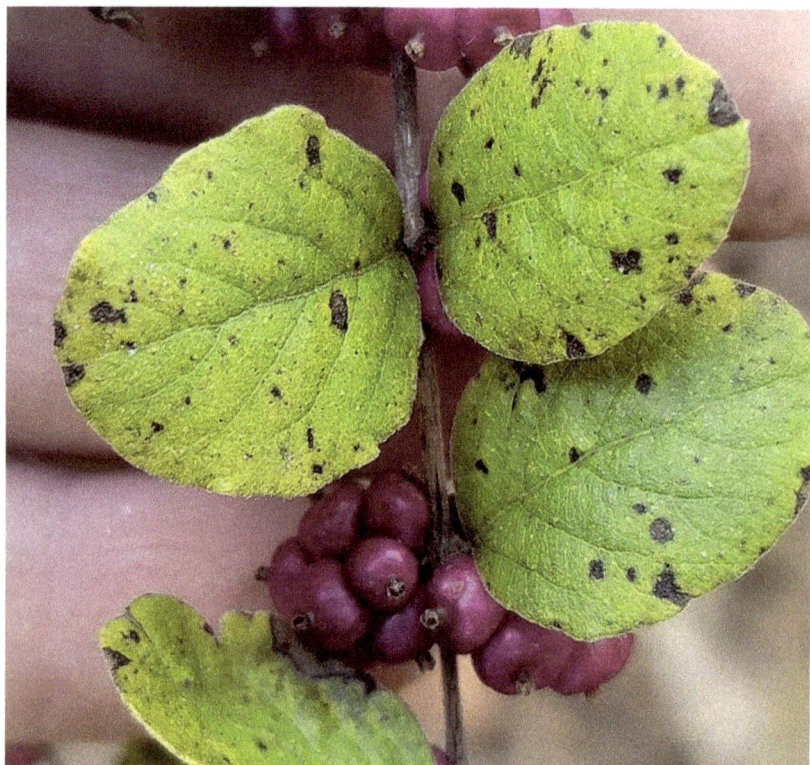

CORALBERRY, *Symphoricarpos orbiculatus*

Distribution Widespread and sometimes common and abundant.
Habitat Pastures, open woods, clearings, and along streambanks or low ground.
Importance Coralberry is an important deer food in the Ozark. The fruit and twigs are eaten consistently during autumn, winter, and spring, and to a lesser extent during summer. The persistent fruits are also eaten during autumn, winter, and spring by ruffed grouse, during autumn and winter by prairie chickens, and occasionally by quail, pheasants, and wild turkeys.

These shrubs are sometimes used in plantings for erosion control, and also provide excellent wildlife cover.

American Basswood, Linden

Description Forest tree, often with several boles clumped together; leaves simple, alternate, broadly ovate or cordate, mostly glabrous, sharply serrate; flowers whitish, fragrant, clustered on a stalk from a leafy bracht, appearing in late spring and early summer; fruit seedlike, pubescent, about 3/16 in. (4–5 mm.) in diameter.

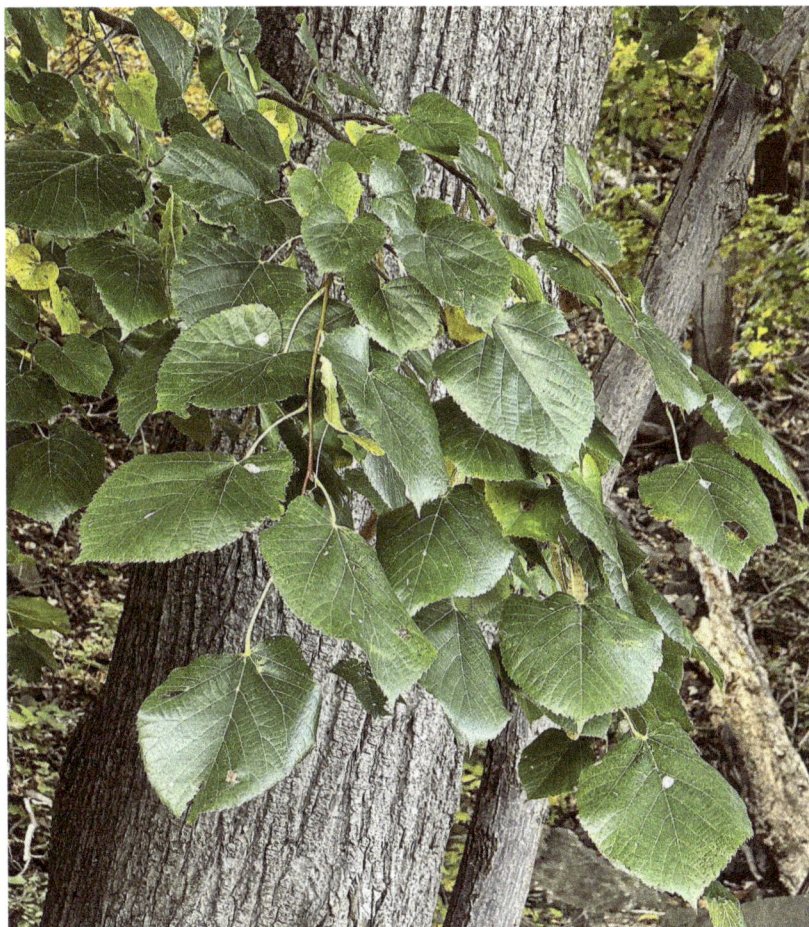

AMERICAN BASSWOOD, *Tilia americana*

Distribution Widely distributed.
Habitat Ravines, lower slopes, bottomlands, and along streams.
Importance Squirrels eat basswood fruits during summer. Basswood may be browsed occasionally when cattle are concentrated in bottoms.

Bees make a flavorful honey from the flowers.

Ulmus spp.

Elm

Description Small to large forest trees; leaves simple, alternate, oblong to obovate, short-petioled, with straight parallel veins from midrib, doubly serrate on margins; flowers small, greenish, in short-stalked clusters, appearing in early spring; fruit a samara, winged, 1-seeded.

AMERICAN ELM, *Ulmus americana*

Distribution Widely distributed.
Habitat On various sites from rocky woods, glades, and ravines, to bottomlands and floodplain.

American elm, *Ulmus americana*, and slippery elm, *Ulmus rubra*, are found in woodlands and floodplains. The first species is identified by smooth buds and leaves, and bark with white and brown layers, the second by fuzzy reddish-brown buds, rough leaves, and dark-brown bark. Slippery elm is more shade tolerant than American elm and occurs more frequently in deep-woods sites.

Winged elm, *Ulmus alata*, is a small sometimes scrubby tree with corky ridges on twigs and small leaves, typical of rocky open woods and glade sites.

All native elms, however, have been affected by Dutch elm disease, a fatal fungal disease spread by bark beetles, and are not as prevalent in Ozark forests are previously.
Importance The elms are important for squirrels; the buds provide essential late winter and early spring food, the seeds are eaten during late spring, and the bark provides emergency winter food. Slippery elm

Elm

SLIPPERY ELM, *Ulmus rubra*

WINGED ELM, *Ulmus alata*
ABOVE, RIGHT

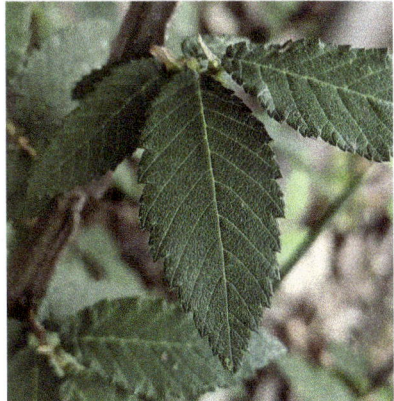

provides some food for ruffed grouse, and the fruits of American elm are occasionally eaten by prairie chickens. Deer browse elm quite heavily when deer numbers are high, but only occasional use is noted with moderate populations and an adequate range.

Cattle browse elm during summer and fall on areas where other forage is scarce.

Deerberry, Blueberry, Sparkleberry

Description Small or large shrubs of the heath family; leaves simple, alternate, oval to oblong-pointed, or obovate, sometimes evergreen, entire on the margins; flowers small, white, greenish to pink, arranged in clusters or solitary, appearing in spring; fruit a small spherical berry, reddish green to dark blue or black, with numerous seeds.

LOW-BUSH BLUEBERRY, *Vaccinium pallidum*

Distribution Widely scattered throughout the region, often abundant.

Habitat Dry oak or pine woods; on acid soils from cherty or igneous rocks.

A most common species is **low-bush blueberry**, *Vaccinium pallidum*, distinguished by low growth, 1–3 ft. (30–90 cm.) high, and firm glabrous leaves with V-shaped base and tapering tip. **Sparkleberry,** *Vaccinium arboreum*, and **deerberry**, *Vaccinium stamineum*, are generally taller, especially the first species, which has thick leathery somewhat evergreen leaves, obovate, with tapering base and rounded tip. The second species, known also as **highbush huckleberry**, is distinguished by deciduous thin-textured leaves, with a pointed apex and a somewhat rounded or cordate base.

Vaccinium spp.

Deerberry, Blueberry, Sparkleberry

SPARKLEBERRY, *Vaccinium arboreum*

**DEERBERRY,
HIGHBUSH HUCKLEBERRY**
Vaccinium stramineum
LEFT, ABOVE

Importance Lowbush blueberry and deerberry leaves, stems, and fruit are eaten by deer throughout most of the year. Grouse obtain small amounts of food from lowbush blueberry during summer. Fruits from these plants are sometimes eaten by quail and turkeys. Studies in other regions indicate that many species of songbirds, especially bluebirds and scarlet tanagers, eat these fruits.

Deerberry and lowbush blueberry fruits are often used in pies and jellies or are eaten fresh.

Blackhaw

Description Small understory trees; leaves simple, opposite, ovate or oblong to obovate, finely serrate; flowers white, with 5 similar petals, in showy clusters, appearing in spring; fruit dark, berrylike, 1-seeded.

RUSTY BLACKHAW
Viburnum rufidulum

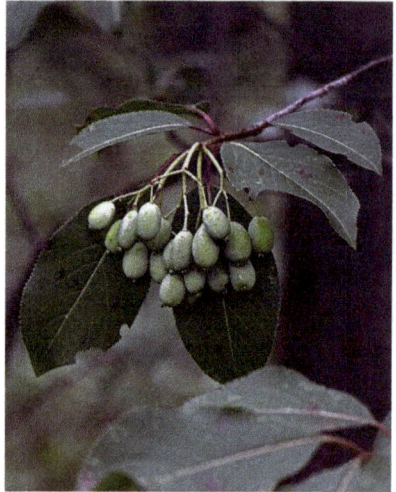

RUSTY BLACKHAW
Viburnum prunifolium
TOP, ABOVE

Distribution Widely distributed.
Habitat Upland woods, thickets, and glades, to low ground and stream bottoms.

An important species is **rusty blackhaw,** *Viburnum rufidulum*, with broadly oval or elliptic leaves, somewhat leathery, glossy, dark green, and with rusty pubescent buds and leaf stalks.

Smooth blackhaw, *Viburnum prunifolium* has thinner pointed, dull leaves and smooth buds and leaf stalks.

Importance Deer eat blackhaw fruits and browse twigs and leaves during summer. The plant also provides food for raccoons and occasionally for ruffed grouse and quail.

Vitis spp.

Grape

Description Woody climbers or bushy spreading vines, with coiling tendrils; leaves simple, alternate, mostly cordate or rounded, long-petioled, variously lobed, with coarsely toothed margins; flowers small, greenish, fragrant, forming clusters in spring; fruit dark blue or black, several-seeded.

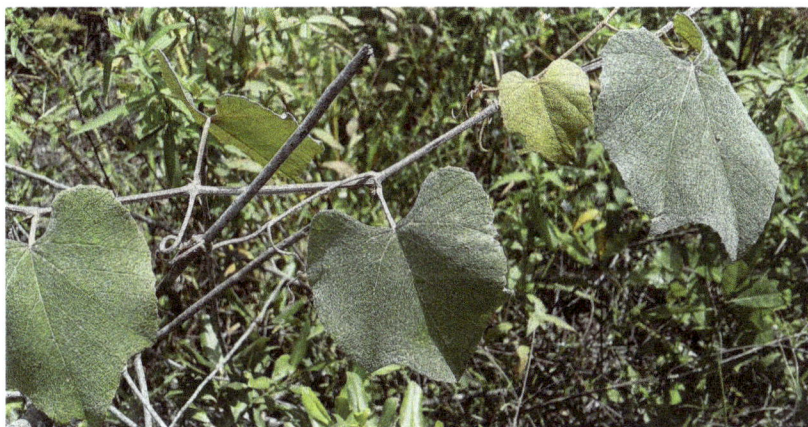

PIGEON GRAPE, *Vitis cinerea*

Distribution Widely distributed.
Habitat Glades, upland woods, streambanks, and bottomlands.

Summer grape, *Vitis aestivalis*, is distinguished by rusty pubescence on undersides of leaves and smooth petioles. **Pigeon** or **sweet winter grape**, *Vitis cinerea*, has ashy gray leaves, and pubescent petioles and is more common on lower ground. **Post oak grape**, a variant of summer grape, is somewhat bushy, with larger fruits than the typical variety and is common in glade sites.

Two other common species are **bushy sand grape**, *Vitis rupestris*, of sandy streambanks, with more or less glabrous foliage and short-pointed orbicular leaves, and **frost grape**, *Vitis vulpina*, which is also

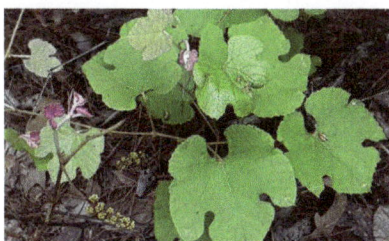

SUMMER GRAPE, *Vitis aestivalis*

glabrous but has more ovate pointed leaves, and high-climbing habit, of low moist woods and alluvial ground.

Muscadine, *Vitis rotundifolia* (synonym *Muscadinia rotundifolia*), is common in the Arkansas Ozarks on low or moist ground. It has rounded or broadly oval leaves similar to sand grape but lacks the shredding bark of the latter.

Grape

BUSHY SAND GRAPE, *Vitis rupestris*

FROST GRAPE, *Vitis vulpina*

MUSCADINE, *Vitis rotundifolia*

Importance *Vitis* probably rates second only to *Quercus* in overall importance to Ozark wildlife. Deer readily (and often excessively) eat *Vitis* during spring and summer. Overbrowsing during summer has been described as an indication of deer overpopulation. Deer also eat the fruits during fall and winter.

Grapes are an important food for ruffed grouse.

Grapes are important to raccoons, wood ducks, wild turkeys, quail, red fox and many nongame species.

When ripe, and especially after a frost or two, these fruits are tasty to humans. They also are used to make wine, preserves, and jellies.

Wildflowers

W ildflowers, sometimes called **forbs**, comprise a group of herbaceous plants, usually with broader leaves and more conspicuous flowers than grass or grasslike species. For convenience, two ferns and a horsetail are included in this section.

Wildflowers make up a major part of the ground vegetation in Ozark forests and ranges. Many perennial wildflowers, especially legumes and composites (Aster family, Asteraceae), are important forage plants for livestock and deer. Some wildflowers have basal rosettes which remain green over winter, and provide valuable green forage when it is otherwise very scarce. Many annual wildflowers are heavy seed producers, and are a major source of food for wild birds. Others are initial invaders on disturbed land and do an effective job of stabilizing bare soil.

Wildlife values are described for each plant or genus in the following pages. However, many plant parts cannot be identified in stomachs or droppings. In wildlife diet studies, this material is genrally grouped together and reported as 'unidentified forbs' or 'miscellaneous materials other than grasses'. In addition to mammals, ruffed grouse, geese and ducks, prairie chickens, wild turkeys, and pheasants also eat significant amounts of miscellaneous leaf material. Quail eat smaller amounts.

GLADE. Woody species are encroaching into this historically more open glade in the Arkansas Ozarks. Prescribed burning can be used to remove the trees and shrubs, while encouraging herbaceous species.

Copperleaf, Three-seeded Mercury

Description Upright annual herbs to 3 ft. (90 cm.) tall; leaves simple, alternate, stipulate; flowers small, inconspicuous, borne in serrated spathelike bracts at base of leaves, appearing in late spring and early summer.

SINGLE-SEED MERCURY
Acalypha monococca

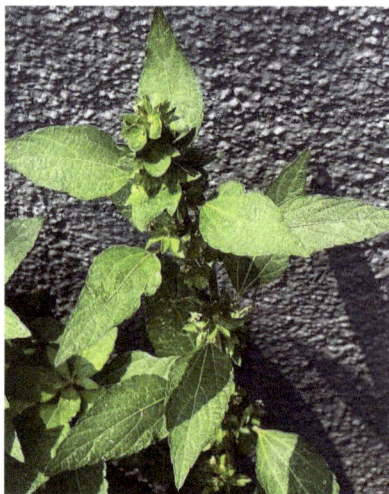

COMMON THREE-SEED MERCURY
Acalypha rhomboidea

VIRGINIA THREE-SEED MERCURY
Acalypha virginica

Distribution Widely scattered throughout the region, seldom abundant.

Habitat Glades, open woods, and waste areas, also low ground and streambanks.

Acalypha monococca has low usually unbranched stems 20 in. (50 cm.) tall or less, and oblong-lanceolate leaves, entire or nearly so.

Two coarser species with serrate leaves are *Acalypha rhomboidea*, with generally smooth stems and ovate leaves, and *Acalypha virginica*, differing by having pubescent stems and more lanceolate leaves.

Importance Doves and, to a lesser extent, quail eat the seeds of these plants. *Acalypha* is poor forage for cattle.

Achillea millefolium

Milfoil, Yarrow

Description Erect perennial herbs with simple, slightly pubescent to woolly stems, usually less than 3 ft. (90 cm.), tall; leaves alternate and also clustered at base, linear to narrow-spatulate or lanceolate, finely dissected; flower heads small, white, in more or less flat-topped to convex clusters, developing in spring and summer.

YARROW, *Achillea millefolium*

Distribution Widespread and sometimes common and abundant.
Habitat Prairies, pastures, and waste ground; naturalized from Europe and Asia.
Importance Yarrow varies greatly in forage value, depending on locality and seasonal development. It provides fair to good forage for sheep and goats, is occasionally eaten by deer, but is rarely grazed by cattle and horses. The flower heads are most often grazed. Its local abundance may be an indicator of continuing or past overstocking. It invades overgrazed areas and increases rapidly.

Yarrow has a very strong odor and, when eaten in large amounts, produces undesirable flavor in milk and milk products.

White Snakeroot

Description Upright perennial herb to 3 ft. (90 cm.) tall with whitish, stringlike roots; leaves simple, opposite, ovate, petioled, serrate, with 3 rather well defined main veins from base of blade; flowers white, in numerous small clusters developing in the latter part of summer.

WHITE SNAKEROOT, *Ageratina altissima*

Synonym *Eupatorium rugosum*
Distribution Widely distributed.
Habitat Moist shaded woods and ravines; on neutral or alkaline soils.
Importance The green plant is poisonous to livestock; dried plants in hay are also toxic but not as dangerous as the fresh state. The poison is soluble in fats or milk and may be transmitted to other animals or humans. In animals, the disease is known as "trembles," in humans as milk sickness. Livestock should not be allowed to graze heavily infested areas. The plants should be pulled and scattered to dry in the sun or sprayed with an herbicide.

Some species are eaten by deer.

Allium spp.

Wild Garlic, Wild Onion

Description Strong-smelling plants of the lily family, with bulbous bases and narrow ascending leaves from ground level or on lower part of stem; flowers or aerial bulblets in terminal umbels, developing in spring, or one species flowering in late summer and fall.

WILD GARLIC, *Allium canadense*

AUTUMN ONION, *Allium stellatum*

Note *Allium* formerly placed in the Lily family, that family now separated into a number of smaller groups.

Distribution Widespread and sometimes common and abundant.

Habitat Open woods, meadows, glades, and waste ground.

Two common wild onions, principally of limestone glades, prairies, and open ground, are **wild garlic**, *Allium canadense*, and **autumn onion**, *Allium stellatum*, both with pinkish flowers, the latter species fall blooming. *Allium canadense* is also found in woods, moist ground, and stream bottoms, and is distinguished by umbels of small aerial bulblets, flowers usually absent, and with all leaves basal.

Another species somewhat less prevalent in the Ozarks is **cultivated garlic**, *Allium sativum*, also with aerial bulblets but with leaves present on the stem. This plant sometimes spreads from cultivation and becomes a nuisance.

Fly poison, formerly of the Lily family (now placed in the Melanthiaceae as *Amianthium muscaetoxicum*) is mentioned because of its somewhat onionlike appearance and high toxicity, especially the bulb. However, in the Ozark range it is not common, and can be distinguished from onions and garlic by its elongated panicle compared to the globular or rounded umbels of *Allium*.

Wild Garlic, Wild Onion

CULTIVATED GARLIC, *Allium sativum*

Importance The wild onions and garlics furnish green succulent forage early in the spring and are readily eaten by most kinds of livestock except horses. Stockmen sometimes turn out livestock early in the spring to use this green forage and frequently damage the better forage species, which are just resuming growth. All onions and garlics produce undesirable flavors in milk and milk products. When dairy animals are the primary users of pastures infested with these plants, grazing should be deferred until later in the season when the plants are less plentiful.

For centuries a medicinal oil has been extracted from *Allium sativum* for bronchitis and nervous diseases of children. The oil has a mild stimulant effect.

Ragweed, Horseweed

Description Upright annual herbs with simple or branching stems; leaves simple, palmately lobed, or dissected, opposite or alternate or sometimes both opposite and alternate on the same plant, mostly sessile or shortpetioled; flowers small, greenish, forming loose terminal spikes, developing in summer and fall.

GIANT RAGWEED, *Ambrosia trifida*

Distribution Widespread and sometimes common and abundant.
Habitat Fields, pastures, glades, waste areas, and low ground.

Giant ragweed, *Ambrosia trifida*, also called **horseweed**, is a coarse plant sometimes reaching 8–10 ft. (2½–3 m.), with opposite leaves, deeply cleft, with 3 to 5 lobes, occurring mostly on deep soils.

Two other ragweeds as much as 3 ft. (90 cm.) tall, but frequently shorter, are common on dry uplands and pastures with thin soils. These are **lance-leaf ragweed**, *Ambrosia bidentata*, identified by simple, alternate, sessile, pointed, lance-shaped leaves, and **common ragweed**, *Ambrosia artemisiifolia*, with both alternate and opposite, mostly dissected leaves.

Ragweed, Horseweed

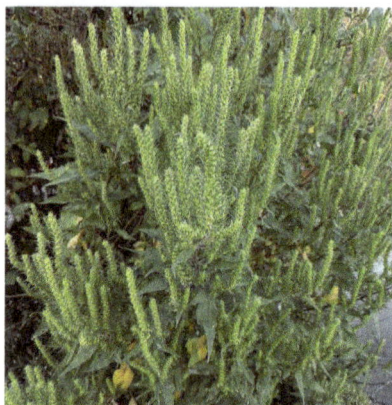

LANCE-LEAF RAGWEED
Ambrosia bidentata

Importance Ragweed seeds are one of the most important quail foods in volume consumed; only Korean lespedeza comes near equaling it in importance. The seeds are also readily taken by doves, prairie chickens, pheasants, and ducks. Turkeys and raccoons use ragweeds only slightly. Deer occasionally eat the plants.

Ragweeds are rated as poor forage for livestock and are seldom eaten unless more palatable forage is lacking. They are considered invaders on native ranges and pastures, but rarely become established Plant and leaf detail. in a vigorous stand of palatable grasses. Ragweed produces undesirable flavors in milk and milk products, and dairy cattle should not be allowed to graze infested areas.

The ragweeds produce large amounts of pollen, causing many cases of hay fever in late-summer and fall.

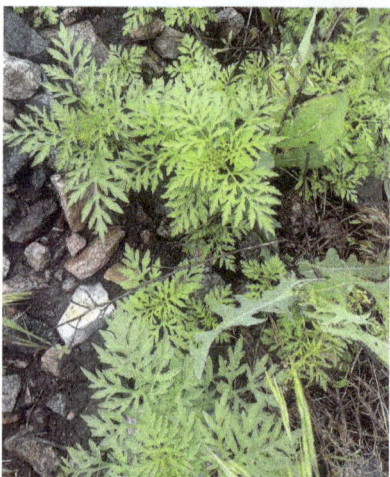

COMMON RAGWEED
Ambrosia artemisiifolia
TOP, ABOVE

Amorpha spp.

False Indigo, Lead Plant

Description Low to tall bushy legumes; leaves alternate, pinnately compound, with 15–35 or more ovate to oblong leaflets with minute dots; flowers small, purplish, in terminal spikelike sprays, developing in summer; fruit small, 1–2 seeded.

LEADPLANT, *Amorpha canscens*

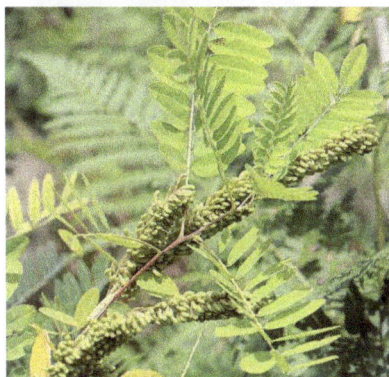

FALSE INDIGO, *Amorpha fruticosa*
TOP, ABOVE

Distribution Widespread throughout the range, except for *Amorpha canescens*, which is uncommon in the southern Ozarks.

Habitat Prairies, glades, open woods to moist bottoms, low woods, and streambanks.

Lead plant, *Amorpha canescens*, of prairies, glades, and upland woods is usually less than 3 ft. (90 cm.) tall, and is distinguished by grayish-pubescent foliage and crowded sessile leaflets.

False indigo or **indigo bush**, *Amorpha fruticosa*, is a much taller shrub of low moist ground and streambanks, with firm green leaflets. It should not be confused with *Baptisia*, another leguminous genus also commonly called indigo but generally found in dry glades and forest openings.

Importance These warm-season, shrublike, native legumes are highly nutritious and provide good forage for all kinds of livestock throughout the year. They decrease on bluestem ranges that are overgrazed, and are sensitive indicators of range condition.

Hog Peanut

Description Trailing or climbing perennial herb; leaves alternate, with 3 ovate or deltoid leaflets, about 2½ in. (5–6 cm.) wide, somewhat pubescent; flowers pale blue in short clusters, developing in summer; aerial pods several-seeded, about 1 in. (2–3 cm.) long. Smaller 1-seeded underground pods are also produced.

HOG PEANUT, *Amphicarpaea bracteata*

Distribution Widely distributed.
Habitat Low woods and shaded sites, mostly on damp ground.
Importance The seeds of this plant make up minor parts of the diet of quail and deer. Wild turkeys scratch out the underground pods in the spring. These pods are often referred to as turkey peas. The plant is considered poor forage for cattle and fair forage for sheep and goats, but is readily taken by hogs in the woods.

Hepatica, Liverleaf

Description Low herb with 3-lobed leaves from the base of plant, remaining green throughout the winter; petioles and flower stalks conspicuously pubescent; flowers with three green bracts and several whitish, pink, or lavender petal-like sepals, appearing in early spring.

HEPATICA, *Anemone americana*

Note Many older references place this species in *Hepatica*, separating plants into several species based primarily on leaf shape.
Synonyms *Hepatica nobilis*, *Hepatica americana*
Distribution Widely scattered throughout the region, seldom abundant.
Habitat Rich woods, ravines, and rocky slopes; on well-drained soils.
Importance Leaves of hepatica are eaten during winter by deer.

Indian Tobacco, Pussytoes

Description Low white-woolly perennial herbs with leafy rosettes and erect flowering stalks, to about 15 in. (30–40 cm.) tall; basal leaves obovate to spatulate, 3-veined, tapering to the petiole, the upper leaves small, bractlike, remote; flower heads whitish, terminal, in small compact clusters, appearing in spring.

INDIAN TOBACCO, *Antennaria plantaginifolia*

Distribution Widely distributed. **Habitat** Open woods, glades, sparse prairies; on dry soils.

The less common *Antennaria neglecta*, more prevalent in the northwestern part of the Ozark range, is identified by narrower basal leaves with one main vein. **Importance** The basal rosettes of pussytoes remain green throughout the winter and provide important winter and early spring wildlife food. Signs of use on these basal leaves are difficult to observe since the entire leaf may be pulled off or the whole plant may be pulled out of the ground. However, stomach samples have shown that a considerable volume is eaten by deer.

Ruffed grouse eat many green leaves during late winter and early spring; the flower heads are eaten along with the leaves in late spring. Observations indicate that green leaves are eaten by turkeys during winter. Prairie chickens also eat this plant, primarily during late winter and early spring.

Pussytoes is poor to fair forage for cattle, sheep, and goats.

Apios americana

Groundnut

Description Trailing or climbing perennial herb, with tuberous rootstocks; leaves alternate, with 3–7 ovate-pointed leaflets, glabrous, with entire margin; flowers purplish brown, in compact clusters, developing in summer; pods narrow, elongate, several-seeded.

GROUNDNUT, *Apios americana*

Distribution Widely distributed.
Habitat Damp woods, meadows, streambanks, and low ground.

Importance This legume is poor forage for cattle and fair for sheep and goats.

Dogbane, Indian Hemp

Description Upright perennial herbs, 2½ ft. (75 cm.) tall or less; stems branching, with smooth, tough, fibrous bark and milky juice; leaves simple, opposite, ovate-oblong to lanceolate, short-petioled, with entire margins; flowers pink or greenish white in axillary and terminal clusters, developing in late spring and summer; fruit consisting of two elongate narrow pods.

DOGBANE
Apocynum androsaemifolium

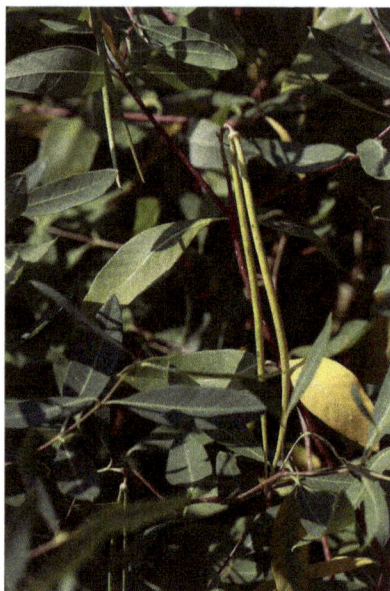

INDIAN HEMP
Apocynum cannabinum

Distribution Widespread and sometimes common and abundant.
Habitat Glades, prairies, open woods, and waste ground.

The two Ozark species are **dogbane**, *Apocynum androsaemifolium*, with pinkish fragrant blooms, and the more common **Indian hemp**, *Apocynum cannabinum*, which has greenish-white flowers.
Importance The dogbanes are worthless as forage for livestock due to a bitter milky juice. The plants are considered poisonous when young and tender to cattle, sheep, and horses, but may also be dangerous when cut and fed in hay. Eradication can be successfully carried out with herbicides.

These plants have a tough fiber bark which has been used as a substitute for hemp. Indian hemp has also been used for medicinal purposes as a cardiac stimulant.

The fragrant flowers attract honeybees for nectar, making a superior, almost colorless honey.

Jack-in-the-Pulpit, Indian Turnip, Green Dragon

Description Perennial herb, 1–2 ft. (30–60 cm.) tall, from a sharp-tasting globose tuber; leaves one or more, large 3-foliate; flowers minute, on a club-shaped spadix in a greenish hooded spathe, in spring; developing a dense cluster of reddish berrylike fruits later in the year.

GREEN DRAGON
Arisaema dracontium

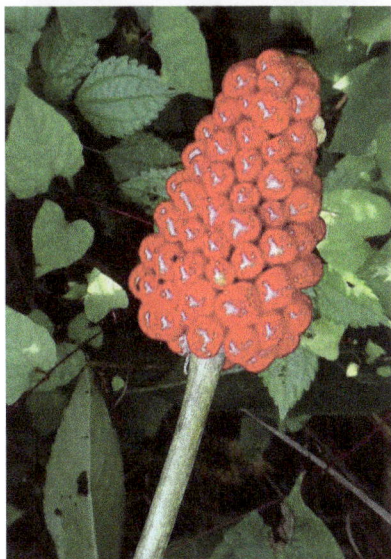

JACK-IN-THE-PULPIT
Arisaema triphyllum

Distribution Widely distributed and probably in every county.
Habitat Low woods and shady ravines; on moist soils.

The **green dragon**, *Arisaema dracontium*, of similar habitats is identified by leaves with more than 5 leaflets and a long-pointed spadix.
Importance Indian turnip is poor forage for cattle, sheep, and goats, but the roots are sometimes eaten by hogs. Most animals usually avoid Indian turnip, probably because of the pungent juice. If leaves or rootstalks are eaten by humans, a violent burning of the tongue and throat results, which may last for several minutes.

The starchy roots of this plant have been cooked and eaten by Native Americans. Boiling, heating, or drying apparently destroys the toxic material. Indian turnip may cause dermatitis to some people.

Wild Ginger

Description Low perennial herb with horizontal rootstock, distinctly aromatic; leaves two, at ground level, broadly cordate or reniform, long-petioled, entire on the margin; flowers solitary, purplish brown, 3-lobed, from base of plant, appearing in early spring.

WILD GINGER, *Asarum canadense*

Distribution Widely scattered throughout the region but seldom abundant.

Habitat Rich woods, lower slopes, and moist ravines.

Importance Wild ginger is poor forage and is seldom eaten by livestock. The fleshy roots are occasionally eaten by hogs. Handling the rootstocks may cause dermatitis in some people.

Asclepias spp.

Butterfly-weed, Milkweed

Description Mostly upright perennials with simple or branching stems and milky juice, to 4½ ft. (1½ m.) tall; leaves simple, opposite, alternate or whorled, linear to broad oblong, short-petioled or sessile, with entire margins; flowers pink, greenish, or reddish orange in umbellate or spherical clusters, developing in late spring and summer; pod splitting on 1 side with numerous silk-tufted seeds.

WHORLED MILKWEED
Asclepias verticillata

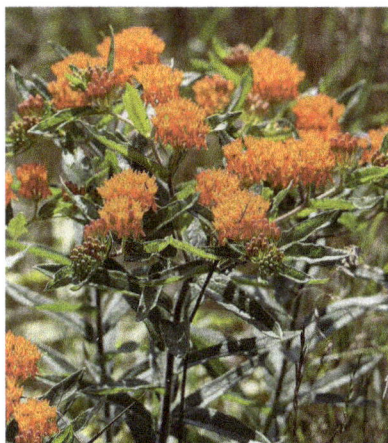

BUTTERFLY-WEED
Asclepias tuberosa

Distribution Generally distributed except for *Asclepias syriaca*, which is less common or absent southwestward.

Habitat Open woods, prairies, glades, fields, roadsides, and waste ground.

Two common species are **whorled milkweed, *Asclepias verticillata***, with usually slender unbranched stems, and narrow leaves in whorls of 3 or more; and **butterfly-weed, *Asclepias tuberosa***, of roadsides and dry soils with branching stems, mostly alternate leaves, and bright orange flowers.

Green-flowered milkweed, *Asclepias viridis*, with some alternate leaves and greenish-purple flowers, occurs in prairies and glades. The **common milkweed, *Asclepias syriaca***, is a coarse thick-stemmed plant with opposite oblong leaves and light purplish-pink flowers.

Butterfly-weed, Milkweed

GREEN-FLOWERED MILKWEED, *Asclepias viridis*

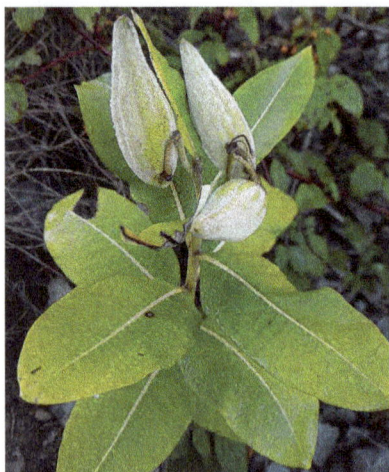

COMMON MILKWEED
Asclepias syriaca

Importance Quail and pheasants occasionally eat milkweed seeds. The milkweeds are worthless to poor livestock forage, taken only when more palatable forage is lacking. Milkweed is considered an invader or increaser on most native ranges. Most milkweed species are generally poisonous to livestock, especially sheep, when eaten in large amounts, although few poisonings have been reported.

Astragalus spp.

Milk Vetch, Ground Plum

Description Perennial branching herbs with a deep taproot; leaves alternate, pinnately compound with numerous leaflets; flowers in spicate racemes, developing in spring and summer; pod thick, plumlike when green, with 2 cavities in cross-section.

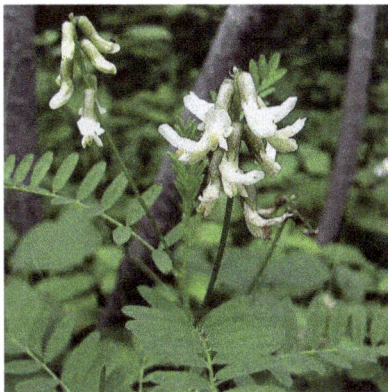

CANADIAN MILK VETCH
Astragalus canadensis

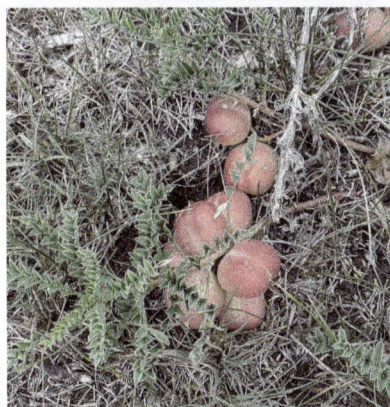

GROUND PLUM
Astragalus crassicarpus
TOP, ABOVE

Distribution Widely scattered throughout the region, seldom abundant.
Habitat Dry prairies, glades, and upland woods.

Ground plum, *Astragalus crassicarpus* var. *trichocalyx*, is a low plant generally less than 20 in. (½ m.) tall, with spreading leaves, and with cream-colored or bluish flowers in spring. *Astragalus canadensis* is a larger plant, as much as 4 ft.

(1.2 m.) tall, with greenish-yellow flowers in dense racemes, appearing somewhat later in spring and into summer.
Importance Although usually high in nutritive content, the milk vetches are seldom eaten by cattle unless more palatable forage is not available. They are sometimes eaten by sheep. Species common to the Ozarks may not be toxic, but they should be suspected if poisoning occurs where these plants are found.

Further west, the green plants and dried tops of several species are known to be poisonous. If eaten over a long period of time, the animals usually become dull, then irregular in gait due to growing weakness, and death sometimes results. Cattle, sheep, and especially horses can be affected.

Aureolaria grandiflora **OROBANCHACEAE, BROOM-RAPE FAMILY**

Gerardia

Description Tall conspicuous perennial, to 3½ ft. (1 m.) tall; leaves simple, mostly opposite becoming alternate on upper stem, lanceolate, coarsely serrate to somewhat pinnatifid, turning black upon drying; flowers conspicuous, yellow, tubular, about 2 in. (5 cm.) long; fruit a dry black many-seeded capsule; not to be confused with mullein foxglove, *Dasistoma macrophylla*, also with yellow flowers but these only about 1 in. (2.5 cm.) long.

GERARDIA, *Aureolaria grandiflora*

Synonym *Gerardia grandiflora* var. *cinerea*
Note Not to be confused with **mullein foxglove**, *Dasistoma macrophylla*, also with yellow flowers but these only about 1 in. (2.5 cm.) long.

Distribution Widely distributed in all but eastern parts of the Ozark region.
Habitat Woodlands, glades, and bluffs; on dry, mostly acid soils.
Importance *Aureolaria* is occasionally eaten by deer, especially when the large flowers are forming.

Baptisia spp.

False Indigo, Wild Indigo

Description Branching perennials, solitary, somewhat bushy, with thick stems up to 3–4 ft. (1–1.2 m.) tall; foliage turning black when dried, the leaves palmately compound with 3 leaflets; flowers showy, blue, whitish, or yellow in elongated racemes, appearing in spring and early summer; pod several-seeded, plump, with narrowing tip, turning black when mature or dried.

Baptisia bracteata

Baptisia alba

Distribution Widespread, except *Baptisia australis* var. *minor*, which is absent from southern portions of the Ozarks (more common westward).

Habitat Prairies, glades, open woods, and low ground.

Blue false indigo, *Baptisia australis*, is a showy plant of the glades, about 3 ft. (90 cm.) tall, with smooth foliage and blue flowers.

Two species more widespread are the low, bushy, somewhat pubescent *Baptisia bracteata* (synonym *Baptisia leucophaea)*, with large leafletlike stipules below the 3 leaflets and heavy racemes of yellow flowers; and the taller more erect

Baptisia alba (synonym *Baptisia lactea*) with whitish flowers in vertical racemes. In the same locality, *Baptisia alba* blooms several weeks later than *Baptisia bracteata*.

Importance The wild indigos provide poor or low-quality forage for cattle, sheep, and goats, being grazed occasionally in the very early spring. Several species are reported to be poisonous to livestock.

Indigo derives its name from past use as a source of a low-quality dye.

Partridge Pea

Description Upright annual herb with few ascending branches about 2 ft. (60 cm.) tall; leaves pinnately compound with 10 or more pairs of linear leaflets, these sensitive and folding to touch, about 1/16 in. (1½–2 mm.) wide, and with oblique base; flowers showy, bright yellow, appearing in summer; pod 1 in. (2½ cm.) long or more, with slight partitions between each seed.

PARTRIDGE PEA, *Chamaecrista fasciculata*

Synonym *Cassia fasciculata*

Distribution Widely scattered throughout the region, occasionally abundant.

Habitat Glades, fields, open woods, and disturbed areas.

Importance Partridge pea is not a dependable food plant for wildlife or livestock because of its sporadic and limited abundance. When available, it provides fair forage for livestock. The seeds make up a small part of the diet of quail and pheasants. Deer occasionally eat the plant.

Chenopodium spp.

Lamb's Quarters, Pigweed, Wormseed

Description Upright or somewhat spreading herbs, mostly annuals, with smooth or mealy branching stems; leaves simple, alternate, ovate to oblong-lanceolate, coarsely dentate or sinuous on margins; flowers minute in greenish spikelike clusters, developing in spring and summer.

LAMB'S QUARTERS
Chenopodium album

WORMSEED
Dysphania anthelmintica

Distribution Widespread and sometimes common and abundant.
Habitat Fields, waste ground, farmsteads, and near buildings; the species included here naturalized from Europe or tropical America.

The widespread **lamb's quarters** or **pigweed**, *Chenopodium album*, from Europe, has coarse stems to 4 ft. (1.2 m.) tall or more, with leaves 1–2 in. (2½–5 cm.) long, whitish or glaucous, somewhat mealy on the surface.

The uncommon **wormseed**, *Dysphania anthelmintica* (synonym *Chenopodium ambrosioides* var. *anthelminticum*), is distinguished by greenish odoriferous foliage, the leaves larger, more elongate or lanceolate. Several other species occur throughout the region.
Importance *Chenopodium* seeds provide a trace amount of food for quail, pheasants, and waterfowl, as well as deer, which also eat the plant. Many species of songbirds are reported to eat the seeds.

Chenopodium species are generally poor forage for livestock and are usually not eaten except in the early spring or if good forage is lacking.

Wormseed is sometimes grown as a crop to obtain the "oil of chenopodium," which is used as an antihelminthic (anti-parasite).

Water Hemlock, Spotted Cowbane

Description Plants tall, to about 6–7 ft. (2 m.), with smooth spotted stems and fleshy tuberous rootstocks; an identifying feature is the chambered stalk at the stem-root junction when cut in vertical section; leaves alternate, compound, with linear or lanceolate serrate leaflets, the lateral veins ending in the notches between the teeth; flowers small, white, in compound umbels, developing in summer.

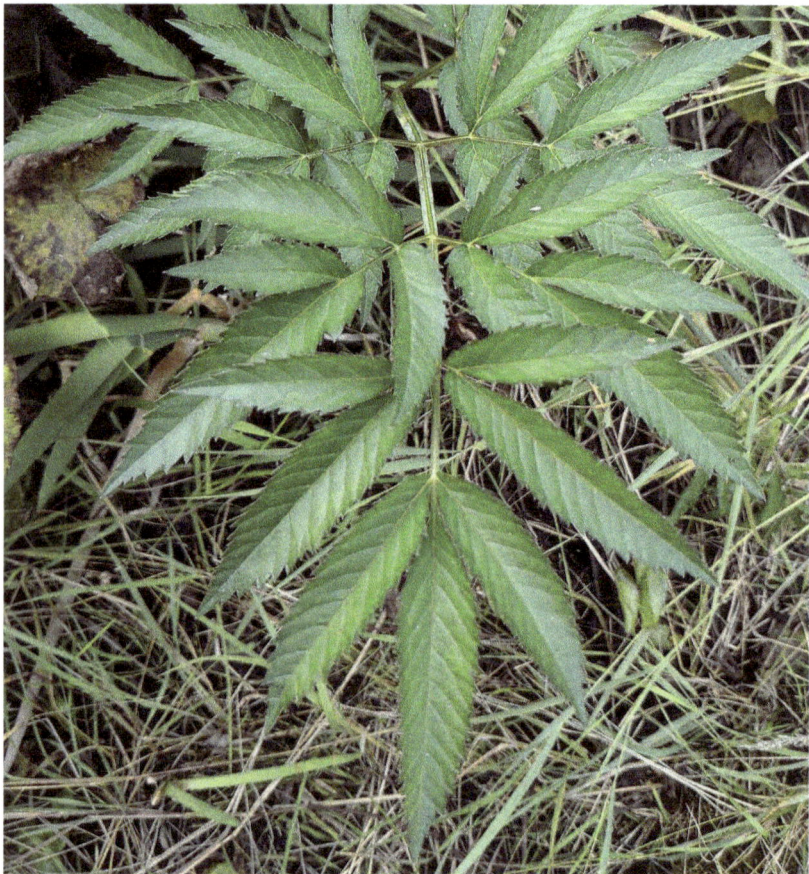

WATER HEMLOCK, *Cicuta maculata*

Distribution Widely distributed.
Habitat Restricted to damp habitats, swales, low ground, and swampy meadows.
Note This species can be confused with **poison hemlock, *Conium*** *maculatum*, a European introduction, and also highly toxic, which has carrotlike foliage, stems splotched with purple, and lacking the tuberous roots (see photos, next page).

Cicuta maculata

Water Hemlock, Spotted Cowbane

WATER HEMLOCK, *Cicuta maculata*

POISON HEMLOCK
Conium maculatum

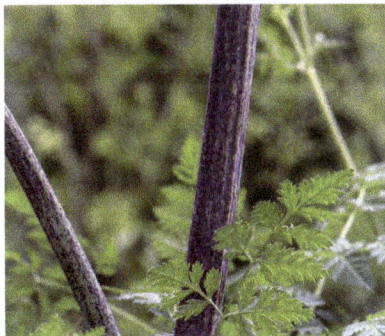

POISON HEMLOCK
Conium maculatum

Importance Water hemlock is abundant in ditches and low meadows and is significant because it is *one of the most toxic of all plants*. The roots and rootstocks have a very distinctive aromatic or musky odor and, when broken, exude an acrid yellowish resinlike substance, the poison cicutoxin.

This plant is poisonous to both livestock and humans. Most livestock losses occur among cattle, but horses and swine are also affected. Children and adults have been killed by eating small amounts of the fleshy roots, which may be mistaken for parsnips. During early growth all parts of the plant are poisonous if eaten, although the root crown and the roots are the most poisonous.

Most livestock poisoning occurs in the early spring when the roots are easily pulled out and eaten along with the green plant parts. The stem and leaves become less poisonous in summer and autumn. The dried seeds and older tops are probably not a source of danger, and the leaves can be eaten safely in hay.

Symptoms of stricken animals include frothing at the mouth, violent convulsions, labored breathing, and evidence of great pain. When severe, death results from respiratory failure.

Heavily infested areas should not be grazed until the plants are completely eradicated.

Butterfly Pea

Description Erect or ascending-trailing perennial herb from a taproot; leaves compound with 3 ovate-elliptic leaflets 1–3 in. (2½–7½ cm.) long; awl-shaped stipels at base of leaflets; flowers large, showy, about 2 in. (5 cm.) long, pale blue or violet with darker striations, appearing in summer; pod linear-oblong, somewhat flattened, about 2 in. (5 cm.) long, and several seeded.

BUTTERFLY PEA, *Clitoria mariana*

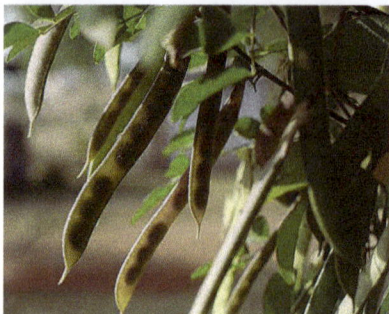

Distribution Southern parts of the region, north to the central Ozarks in Missouri.

Habitat Woodlands, glades, or bottomlands; on dry acid soils.

Importance Butterfly pea is of limited use for wildlife or livestock because of its limited occurence.

Coreopsis spp.

Tickseed

Description Annual and perennial herbs of variable height, stems of most species glabrous; leaves opposite, simple, sometimes deeply cleft, or leaflets distinct, with smooth margins, short-petioled or sessile; flower heads mostly yellow, the rays with broad notched tips developing in spring and early summer.

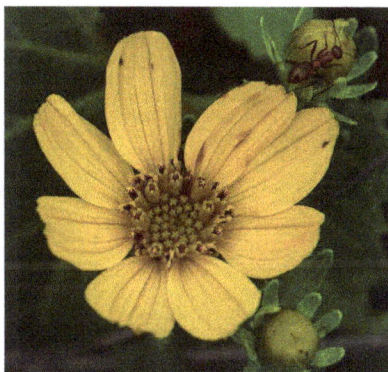

STIFF TICKSEED, *Coreposis palmata*

Distribution Widely scattered throughout the region, seldom abundant.

Habitat Dry woods, prairies, glades, and open ground.

Coreopsis palmata is a common species, a glabrous perennial with sessile 3-lobed leaves, the segments narrow-linear.

Two other perennials are **lance-leaf tickseed**, *Coreopsis lanceolata*, with simple glabrous mostly unlobed leaves, and **tall tickseed**, *Coreopsis tripteris*, a conspicuous plant with coarse stems 3–6 ft. (1–2 m.) tall or even more, and with petioled compound leaves consisting of 3–5 glabrous leaflets.

Coreopsis grandiflora is also a perennial with finely parted leaves, each with 3–5 divisions, and a long leafless flower stalk. The leaves on the lower stem are usually undivided.

Coreopsis tinctoria has bright yellow ray flowers with purplish base, and is the only annual species identified by pinnately divided leaves with narrow elongated lobes.

Importance Seeds (achenes) of *Coreopsis* are eaten by prairie chickens and turkeys. Deer make only occasional use of this plant.

The tickseeds rate as poor forage for cattle and fair for sheep, usually eaten only in early spring when more palatable forage is scarce.

Tickseed

LANCE-LEAF TICKSEED *Coreopsis lanceolata*

LARGE-FLOWER TICKSEED
Coreopsis grandiflora

TALL TICKSEED
Coreopsis tripteris

GOLDEN TICKSEED
Coreopsis tinctoria

Croton

Description Annual herbs, strong-smelling, with scurfy rough-pubescent or woolly stems, to about 10–12 in. (25–30 cm.) tall or more; leaves simple, alternate, entire or sometimes toothed on the margin; flowers small, inconspicuous, appearing in summer and fall.

HOGWORT, *Croton capitatus*

Distribution Widespread and sometimes common and abundant.

Habitat Fields, pastures, glades, overgrazed rangeland, and waste areas; on dry soils.

Two common species are **hogwort**, *Croton capitatus*, woolly pubescent, the leaves about 1½ in. (4 cm.) long with entire margins, and **prairie-tea**, *Croton monanthogynus*, somewhat scurfy, with smaller leaves mostly less than 1½ in. (4 cm.) long.

A third species and not as widespread is *Croton glandulosus*, distinguished from the above species by serrate leaves.

Importance The hard-coated croton seeds are eaten throughout year by doves, quail, turkeys, and prairie chickens.

Crotons are worthless as forage for livestock, and seldom eaten unless more palatable forage is lacking. On ranges and pastures they are generally an indicator of overgrazing and trampling.

Cattle and sheep may be poisoned by eating large amounts of these plants, either green or in hay. Few occurrences of poisoning have been reported, however, since animals seldom eat this plant, probably because of the disagreeable taste and strong scent. Milk goats apparently become addicted to ripening croton seeds which, consumed in large amounts, have purgative properties.

The nectar reportedly is toxic to honey bees. Honey from poisoned bees may cause sickness in humans.

Croton

PRAIRIE-TEA, *Croton monanthogynus*

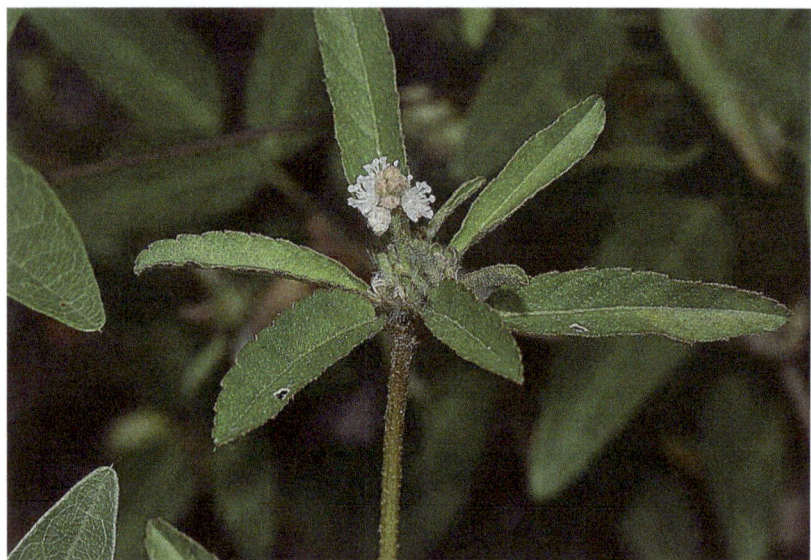

Croton glandulosus

Croton willdenowii

Rushfoil

Description Delicate annual plant; stems slender, scale-covered, to about 15 in. (40 cm.) tall; leaves simple, appearing opposite, linear-elliptic, sessile or nearly so, silvery-scurfy beneath, and with entire margins; flowers inconspicuous, in small clusters, developing in summer; fruit dry, 1-seeded.

RUSHFOIL, *Croton willdenowii*

Synonym *Crotonopsis elliptica*
Distribution Widely distributed.
Habitat Rocky glades, fields, and dry woods; on acid soils.

Importance Rushfoil seeds are eaten by quail and occasionally by deer and prairie chickens.

Cunila origanoides   LAMIACEAE, MINT FAMILY

Dittany

Description Low perennial herb, tufted, with smooth branching stems, square in cross section, about 10 in. (25 cm.) tall; leaves aromatic, simple, opposite, ovate-pointed, glabrous, serrate, with short petioles; flowers small, whitish purple, appearing in middle to late summer.

DITTANY, *Cunila origanoides*

Distribution Widely scattered throughout the region, seldom abundant.
Habitat Dry wooded sites, on cherty

or sandy acid soils.
Importance Dittany provides poor to fair forage for cattle, but is better forage for sheep and goats.

Dalea spp.

Prairie Clover

Description Upright perennial herbs as much as 3 ft. (90 cm.) tall; leaves alternate, compound with 3–7 narrow-linear to oblong leaflets, entire on the margin; flowers small, in dense cylindrical heads, roseate or white, developing in late spring and summer; fruit minute, 1 or 2 seeded.

PURPLE PRAIRIE CLOVER
Dalea purpurea

WHITE PRAIRIE CLOVER
Dalea candida

Synonym *Petalostemon* spp.
Distribution Widely distributed.
Habitat Prairies, glades, open woods, and bluffs.

Purple prairie clover, *Dalea purpurea* (synonym *Petalostemon purpureus*), is distinguished by leaves with 3–5 linear leaflets about 1/16 in. (1½–2 mm.) wide and dense heads of roseate flowers. **White prairie clover, *Dalea candida*** (synonym *Petalostemon candidus*), has 5–7 oblong leaflets up to ⅜ in. (9–10 mm.) wide and white flowers.

Importance Prairie clover is an excellent forage plant and is readily eaten by all kinds of livestock as well as deer and turkey. Purple prairie clover is highly nutritious and has a high protein content, particularly in the new growth. The prairie clovers are considered decreasers where ranges are overgrazed, but are usually not a good indicator of range condition because they are not abundant.

Jimson Weed

Description Coarse, branching annual with unpleasant smell; stems thick, glabrous, to about 3 ft. (90 cm.) tall; leaves simple, alternate, dark green, ovate, coarsely toothed; flowers large, white, tubular, 3–4 in. (8–10 cm.) long, developing in summer; fruit a tough prickly capsule with numerous seeds.

JIMSON WEED, *Datura stramonium*

Distribution Widespread and sometimes common and abundant.
Habitat Fields, pastures, and waste areas; an introduced weedy species from Asia.
Importance Jimson weed is seldom taken by cattle but is sometimes eaten by sheep and goats. The rank heavy scent may be responsible for its unpalatability. Jimson weed leaves, seeds, and roots are toxic to livestock, although poisoning is seldom fatal. Seeds are occasionally eaten by pheasants.

Poisonings in humans have been reported from eating unripe seed pods. Contact with flowers and foliage may also cause dermatitis to some people.

Queen Anne's Lace, Wild Carrot

Description Erect biennial herb from a taproot; stems pubescent to 4–5 ft. (1.2–1.5 m.) tall; leaves alternate, compound, dissected, with numerous small divisions, the petiole V-shaped at its base; flowers white, small, in flat-topped umbels, usually a single dark flower at center, developing in spring and throughout the summer.

QUEEN ANNE'S LACE, *Daucus carota*

Distribution Widespread and sometimes common and abundant.
Habitat Fields, roadsides, and waste ground; naturalized from Eurasia.
Importance Wild carrot is poor to fair forage for livestock and is usually eaten in quantity only when more palatable forage is not available. When large amounts are eaten by cows, the milk becomes tainted with a bitter flavor.

The leaves, especially when wet, can produce a dermatitis in some people.

Larkspur, Staggerweed

Description Upright perennial herbs, the coarser species to 3 ft. (90 cm.) tall; leaves alternate or basal, palmately lobed or deeply cut, with long petioles on lower stem becoming shorter upward; flowers with 1 spur, bluish, purple, or white, arranged in terminal racemes, in spring.

DWARF LARKSPUR
Delphinium tricorne

CAROLINA LARKSPUR
Delphinium tricorne

Distribution Widespread, or some species with limited range.
Habitat Glades, prairies, open woods, and stream bottoms.

Dwarf larkspur, *Delphinium tricorne*, a common species, has short soft stems, usually less than 20 in. (50 cm.) tall, with blue or white flowers in short racemes.

Two taller species are *Delphinium treleasei* with long-stalked flowers, and *Delphinium carolinianum* with narrow racemes of nearly sessile flowers. *Delphinium treleasei* is uncommon, and most prevalent in the limestone glades and barrens of the western Ozarks.

Importance Larkspurs are poisonous to cattle. Most cases of toxicity are reported from eating the numerous species of larkspurs that occur in the plains and mountain ranges, although the eastern species including those in the Ozarks are also considered toxic. Most poisoning is in the spring when the cattle will occasionally eat the green plants. The entire plant contains several toxic alkaloids. Sheep are rarely affected under range conditions. Cattle should not be allowed to graze heavily infested areas, especially in the early spring.

Desmodium spp.

Beggar's Ticks, Tick Clover, Tick Trefoil

Description Perennial herbs with trailing or erect habit, some species to 3 ft. (90 cm.) tall; leaves alternate, compound with 3 leaflets, each with minute needlelike stipules at base; flowers mostly roseate purplish, in loose axillary or terminal panicles, developing in summer; pod several-seeded, articulated, breaking into sections and sticking to hair and clothing.

Desmodium rotundifolium

Desmodium nudiflorum

Distribution Widely scattered throughout the region; occasionally abundant.

Habitat Primarily dry open woods on cherty, sandy, acid soils, and less commonly on low ground and in shaded places.

Several important or common species occur. *Desmodium rotundifolium* is trailing or procumbent, with rounded leaflets and pubescent stems; it is a common legume of dry woods in the Ozarks. In similar sites, species with upright stems, mostly branching, include *Desmodium marilandicum* with small ovate leaflets about ½ in. (10–12 mm.) wide and elongate leafy flowering racemes; *Desmodium nudiflorum* (synonym *Hylodesmum nudiflorum*) with larger leaflets clustered at top of main stem and with leafless unbranched flowering stalks arising from the base; and *Desmodium paniculatum* with more elongate or lanceolate leaflets to 3–4 in. (7½–10 cm.) long.

Pointed-leaf Tick-clover, *Desmodium glutinosum* (synonym *Hylodesmum glutinosum*) of more shaded sites has leaves clustered at midstem below the flowering stalk, the individual leaflets large, somewhat rounded or broadly ovate, with attenuate point.

Beggar's Ticks, Tick Clover, Tick Trefoil

Desmodium paniculatum

Desmodium glutinosum

Several other trefoils also widespread in the Ozarks include ***Desmodium ciliare*** with nearly sessile leaves and oval-oblong leaflets; ***Desmodium cuspidatum*** with long-petioled leaves and acuminate leaflets, and large stipules to 2/5 in. (1 cm.) long; and ***Desmodium obtusum*** with conspicuous foliaceous stipules about 1/5 in. (5 mm.) long and nearly sessile leaves and ovate leaflets. ***Desmodium sessilifolium*** is identified by sessile leaves with narrow linear leaflets. Other species also occur.

Importance These plants rank among the most important year-round ruffed grouse foods in Missouri. Turkeys also eat considerable quantities of these plants. Quail make heavy use of the seeds, and pheasants eat them occasionally. Tick trefoils are considered a preferred food for deer in the Ozarks.

The tick trefoils provide fair to very good forage for livestock. Like most legumes, they generally have a high protein and calcium content.

Most trefoils will decrease under heavy grazing but are usually not present in sufficient amounts to be a good indicator of range condition.

Dicentra spp.

Dutchman's Breeches, Squirrel Corn

Description Low herbaceous perennials with a cluster of small bulbs or tubers, more or less stemless; leaves finely dissected, long-petioled, pale green, from base of plant; flowers pinkish or cream colored, V-shaped, forming nodding racemes, developing in spring.

SQUIRREL CORN
Dicentra canadensis

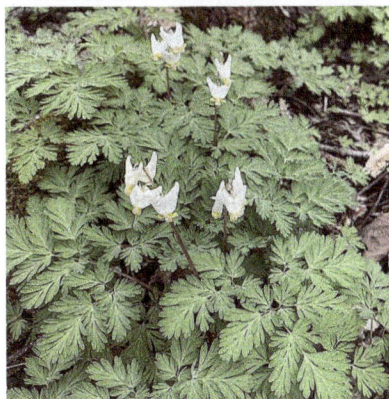

DUTCHMAN'S BREECHES
Dicentra cucullaria
TOP, ABOVE

Distribution Widespread and sometimes locally abundant.

Habitat Upland woods, ravines, and stream bottoms.

Dutchman's breeches, *Dicentra cucullaria*, is common with whitish-pink flowers, distinguished from the less abundant **squirrel corn**, *Dicentra canadensis*, by the latter's cream-colored flowers and less divergent floral spurs.

Importance These plants are poor forage, eaten only if suitable forage is not available. Some species are poisonous to livestock, but Dutchman's breeches is probably the most dangerous. Most poisoning in cattle occurs in early to mid-spring when woodland pastures are being grazed. Fatal poisonings have been reported, but the plants are generally not eaten in large amounts.

Diodella teres

Diodella teres RUBIACEAE, MADDER FAMILY

Buttonweed, Poor Joe

Description Branching annual with spreading stems to 20 in. (50 cm.) tall; leaves simple, opposite, linear-lanceolate, sessile, about 1 in. (2.5 cm.) long; stipules bristlelike; flowers small, whitish or pinkish, clustered in axils of the leaves, appearing in summer; fruit small, obovoid, breaking into two indehiscent parts.

BUTTONWEED, *Diodella teres*

Synonyms *Diodia teres, Hexasepalum teres*
Distribution Widely distributed.
Habitat Dry glades, barrens, waste areas, and alluvial ground; on sandy, rocky, or sterile soils.

Importance Buttonweed provides food for wild turkeys, quail, and prairie chickens during most of the year. Deer occasionally eat this plant.

Coneflower

Description Upright perennial herbs; stems simple, mostly unbranched, to 3 ft. (90 cm.) tall; leaves simple, alternate, linear-lanceolate, or more broadly ovate with tapering petioles on lower stems; flower heads with drooping purplish or yellow rays, solitary, on terminal stalks, developing in summer.

PURPLE CONEFLOWER
Echinacea purpurea

Echinacea paradoxa

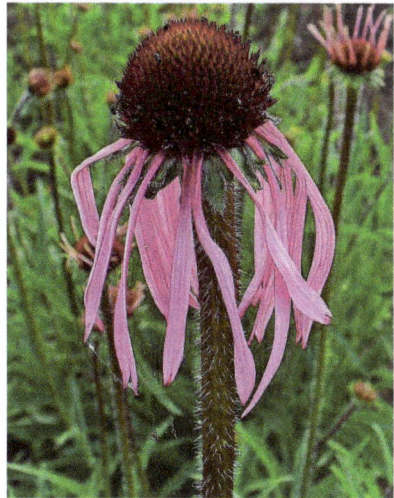

Echinacea pallida

Distribution Scattered throughout the region, occasionally abundant. **Habitat** Glades, prairies, dry woods, and low ground.

The widespread **purple coneflower**, *Echinacea purpurea*, is identified by ovate leaves on lower stem, and drooping purple rays.

Two coneflowers with distinctively elongate linear or lanceolate leaves with tapering petioles on the lower stem are *Echinacea pallida*, with light purplish, sometimes white, ray flowers and pubescent foliage, also common; and *Echinacea paradoxa* with yellow rays. and most common on calcareous soil in the western Ozarks

Importance The coneflowers are highly palatable and provide good forage for all kinds of livestock. They have a high nutritive content, especially in early spring. The plants are decreasers in native ranges and soon disappear with continued close grazing, giving way to less palatable plants.

Horsetail, Scouring Rush

Description Leafless plants, the erect jointed stems, hollow, simple, or with whorls of slender scaly branches, to 3 ft. (90 cm.) tall; fruiting structure terminal, conelike, producing spores.

SCOURING RUSH
Equisetum hyemale

Habitat Low ground, streambanks, sandy alluvium, waste areas.

Field horsetail, *Equisetum arvense* is a short-lived perennial, the sterile stems with spreading whorled branches not more than 3 ft. (90 cm.) tall.

Scouring rush, *Equisetum hyemale* is a coarse perennial frequently taller than 3 ft., mostly simple-stemmed or with a few irregular branches, remaining green in winter.

Importance *Equisetum* is poisonous to all kinds of livestock throughout the year and is seldom eaten except in hay. Cattle, horses, and sheep have been poisoned by eating heavily infested hay. *Equisetum* plants are particularly difficult to eradicate from rangelands.

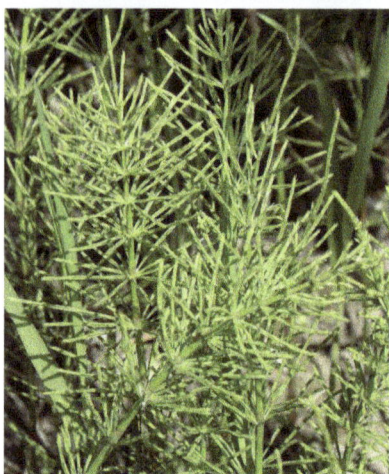

FIELD HORSETAIL
Equisetum arvense
TOP fertile stems, LOWER sterile stems

Distribution Widely scattered, sometimes locally common and forming dense stands.

Burnweed

Description Tall conspicuous annual to about 6–7 ft. (2 m.) tall, the stem ridged or grooved; leaves simple, alternate, oblong-lanceolate, coarsely toothed; flowers small, whitish, in compact heads, appearing in middle to late summer.

BURNWEED, *Erechtites hieraciifolius*

BURNWEED, *Erechtites hieraciifolius*

Distribution Widely scattered throughout the region, sometimes locally abundant.
Habitat Open woods, roadsides, waste ground, frequently occurring on burned areas.

Importance Burnweed is an early invader on disturbed areas and provides protection against erosion until perennials become established. It is sometimes grazed by cattle for a short period during early summer.

Fleabane, Horseweed

Description Erect annuals or short-lived perennials with leafy simple or branching stems, as much as 6–7 ft. (2 m.) tall; leaves simple, alternate, linear or lanceolate to obovate, mostly sessile but sometimes petioled on lower stem; flower heads with white, pink, or purple rays surrounding a yellow center, and subtended by a simple involucre consisting of a single whorl of greenish bracts, developing in spring and summer; the fleabanes appear similar to *Aster* spp., but the latter are all perennial, late-blooming, and have an involucre of several whorls of imbricated or overlapping brachts below the ray flowers.

Erigeron pulchellus

Distribution Widely distributed. **Habitat** Glades, prairies, open woods, fields, and waste areas.

A common fleabane is *Erigeron pulchellus*, with a somewhat tufted rosette of spatulate or obovate leaves, those on the stem few, more linear, and smaller, and with white,

pink, pale blue, or pale purple ray flowers.

Other species and closely related genera include the **whitetop** or **daisy fleabanes**, *Erigeron annuus* and *Erigeron strigosus*, with white ray flowers, of fields, clearings, and waste ground; and **dwarf** or

Erigeron spp.

Fleabane, Horseweed

Erigeron strigosus

HORSEWEED, *Erigeron canadensis*
ABOVE, LEFT (leaf detail)

spreading fleabane, *Erigeron divaricatus* (synonym *Conyza ramosissima*), of similar sites but with bushy branching habit, about 10–12 in. (30 cm.) tall, and with purplish ray flowers.

The common **horseweed**, *Erigeron canadensis* (synonym *Conyza canadensis*), is a tall leafy plant, branching toward the top into a large elongate inflorescence, with numerous minute whitish ray flowers.

Importance Species of *Erigeron* vary from poor to fair forage for livestock, being more palatable to sheep and goats than cattle. They are eaten to some extent in the early stages of growth but seldom later on in the season. Whitetop fleabane increases in heavily grazed pastures and is considered an indicator of range abuse. The common horseweed is suspected of being poisonous and is even less palatable than whitetop fleabane. *Erigeron* is eaten by deer occasionally in the spring and summer on poor condition deer ranges.

Horseweed can also cause dermatitis in some individuals.

Rattlesnake Master, Yuccaleaf Eryngo

Description Upright perennial herb of the parsley family with stiff coarse stems to 4 ft. (1.1 m.) tall; leaves elongate, with parallel veins and spiny margins; flowers whitish, in prickly spherical clusters about 1 in. (2½ cm.) in diameter, from stiffly ascending terminal branches, developing in summer.

RATTLESNAKE MASTER, *Eryngium yuccifolium*

Distribution Widely scattered throughout the region, seldom abundant.

Habitat Glades, prairies, upland woods.

Importance This is an important forb of the tall-grass prairie, and livestock relish the nutritious new growth. It disappears along with the better grasses as the range is overgrazed.

These plants have had some medicinal value historically; extracts from the roots were used for treating diseases of the liver.

Euphorbia spp.

Spurge

Description A large genus growing worldwide, represented in the Ozarks by several herbaceous species of various habits from prostrate spreading annuals to coarse perennials, all with milky juice in the stems and leaves; leaves simple, alternate or opposite, dentate or serrate to nearly smooth on margins; true flowers small, inconspicuous, developing a distinctive 3-lobed fruit.

Euphorbia dentata

FLOWERING SPURGE
Euphorbia corollata

Distribution Widely distributed
Habitat Glades, prairies, open woods, fields, and waste areas, on dry soils; also low ground and alluvium.

Common in glades, prairies, and dry woods is **flowering spurge**, *Euphorbia corollata*, a deep-rooted perennial with glabrous stems to 3 ft. (90 cm.) tall or more; leaves alternate, oblong, sessile, and more or less entire on margins; petallike bracts, white, 5-parted, in showy terminal umbellate or spreading inflorescences.

Wood spurge, *Euphorbia commutata*, is an upright annual species of low woods and ravines with alternate sessile leaves entire on the margin and opposite bractlike leaves below the flowering branches.

Two other erect spurges with dentate or lobed petioled leaves are *Euphorbia dentata* and *Euphorbia heterophylla*. The first species is distinguished by toothed, ovate-lanceolate, mostly opposite leaves, the second by variously lobed to unlobed alternate leaves, the bases often marked with red.

Two common annuals of open ground and fields with opposite leaves are **milk purslane**, *Euphorbia supina*, with prostrate matlike habit, much-branched, and small oblong leaves about ½ in. (12–15 cm.) long or less; and, **nodding spurge**, *Euphorbia maculata*, with upright habit and leaves larger, to 1 in. (2½ cm.) long or more. Other less common species also occur.

Spurge

Euphorbia heterophylla

NODDING SPURGE
Euphorbia maculata

MILK PURSLANE, *Euphorbia supina*

Importance Seeds of several species of spurges, notably *Euphorbia heterophylla, E. supina, E. maculata, E. obtusata*, and *E. corollata*, are eaten by mourning doves, turkeys, quail, pheasants, and prairie chickens. Deer eat small amounts of *Euphorbia maculata, E. heterophylla, E. corollata*, and *E. dentata*.

Spurges are worthless to poor forage for livestock, eaten only when other forage is not available. Spurges commonly increase on overused ranges as the better forage plants decrease. The spurges are considered poisonous to livestock and, although they are seldom eaten by choice, they may be consumed in hay and can be fatal.

The plants contain a milky juice, which causes dermatitis in some people.

Galactia volubilis
Milk Pea

Description Low trailing or twining herb; leaves trifoliate, alternate, with ovate-elliptic leaflets, ⅜ in. (1 cm.) wide, rounded at tip, entire on the margin; flowers pink, several, in narrow racemes, developing in summer; pods pubescent, several-seeded, 1–2 in. (2½–5 cm.) long.

MILK PEA, *Galactia volubilis*

Distribution Widely scattered throughout the region, seldom abundant.
Habitat Open woods and glades; on dry cherty or sandy soils.

Importance Milk pea seeds are relatively important in the diet of quail. The plant is occasionally taken by deer. It is a poor to fair forage plant for livestock, being more palatable to sheep and goats than cattle.

Bedstraw, Cleavers, Goose Grass

Description Low or sprawling annuals or perennials with soft 4-angled stems; leaves simple, opposite or whorled, sessile; flowers small, white or purplish, with 5 petals developing in spring and summer; fruit spherical, seedlike, in 2's.

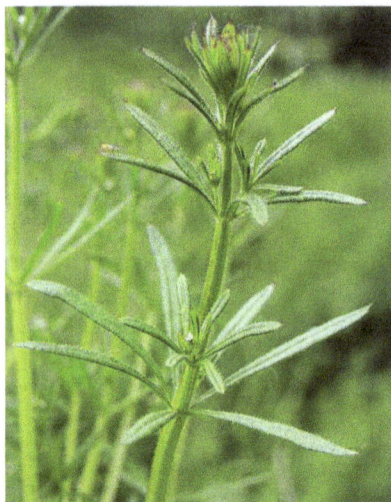

ANNUAL GOOSE GRASS
Galium aparine

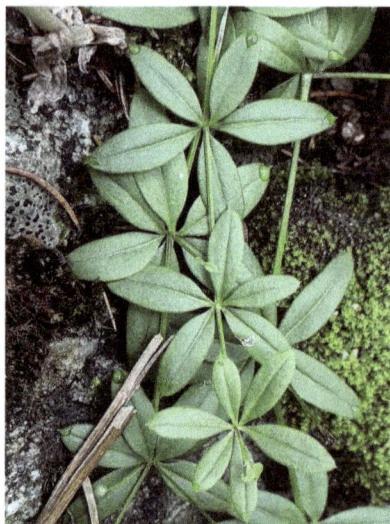

Galium triflorum

Distribution Widely scattered throughout the region.

Habitat Glades and woodland sites on dry to moist ground, also waste areas, stream bottoms, and alluvium.

Several species are common in the region. These include **annual goose grass**, *Galium aparine*, with sprawling habit, harsh stems, and bristly fruits, of waste ground and fields on deep soils.

Two perennials are *Galium triflorum* and *Galium circaezans* also with bristly fruits, but more commonly found in native rich woods locations.

Perennials with smooth fruits are *Galium arkansanum* with purplish flowers, restricted to dry Ozark habits, and the more wide-ranging *Galium concinnum* with white flowers. Several other species of varying distribution and site also occur.

Importance The fruits of *Galium* provide good food for turkeys and are eaten in small amounts by pheasants, deer, ruffed grouse, and prairie chickens. These plants are only fair forage for sheep and goats and worthless to poor for cattle and horses. They will increase with overgrazing.

Several species of bedstraw were used medicinally during early settlement of this country. Fragrant bedstraw and some others were also used at that time as dyes.

Galium spp.

Bedstraw, Cleavers, Goose Grass

Galium arkansanum

Galium circaezans

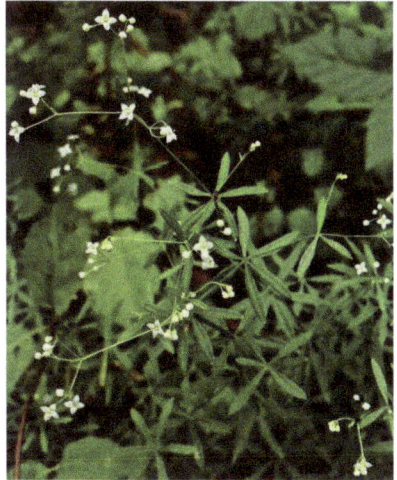

Galium concinnum

Bitterweed, Sneezeweed

Description Erect annual or perennial herbs with simple or branching stems, to 3 ft. (90 cm.) tall; leaves simple, alternate, sessile, narrow, filiform or linear-lanceolate to elliptic; flower heads on terminal stalks with yellow spreading rays surrounding a globose center, the tips of the ray flowers somewhat cleft or 3-lobed, developing in late spring to autumn.

BITTERWEED, *Helenium amarum*

Helenium autumnale

Distribution Widespread and sometimes common and abundant.
Habitat Dry glades, pastures, prairies, swales, swampy ground, and alluvium.

Bitterweed, *Helenium amarum*, is a low annual about 20 in. (50 cm.) tall with leafy stems and narrow filiform leaves 1/16 in. (1½ mm.) wide, found usually in pastures and waste ground.

Two perennial species of moist ground, prairie swales, and streambanks, with wider elliptic-lanceolate leaves are *Helenium flexuosum* and *Helenium autumnale*, the stems noticeably winged with decurrent leaf blades. *Helenium flexuosum* is distinguished by the purplish-brown center, and *Helenium autumnale* by the yellow center, of the flower head.

Importance Livestock seldom eat *Helenium* by choice. It is considered a pest on low-vigor ranges and often becomes a troublesome weed. The plants are toxic if eaten in sufficient quantity. Sheep, cattle, and horses have been poisoned by eating large amounts of the plant, especially the seed heads. *Helenium* also causes a bitter flavor in milk and milk products when eaten in large amounts.

Helianthus spp.

Sunflower

Description Species included here are all upright perennials, mostly rhizomatous. Several species are coarse and tall, to 6–8 ft. (1.8–2.4 m.) or more; leaves simple, opposite, subopposite, or alternate, sessile or petiolate, with 2 more or less defined lateral veins from base of blade, mostly serrate or sometimes entire on the margin; flower heads with conspicuous yellow obtusely pointed rays surrounding a flat dislike receptacle, developing in summer.

Helianthus occidentalis

Distribution Widespread and sometimes common and abundant.
Habitat Glades, prairies, open woods, waste areas. low ground, and streambanks.

Helianthus occidentalis is distinctive, with glabrous purplish stems and mostly basal ovate-lanceolate leaves tapering to long petiole. *Helianthus mollis* and *Helianthus hirsutus* are distinguished by pubescent stems and sessile or nearly sessile opposite leaves, the first species with broadly ovate whitish-villous leaves, cordate at base, and the second species with ovate-lanceolate leaves, not whitish, and with less rounded base.

Two species with at least some opposite leaves on lower stem, becoming alternate upward, and with distinct petioles are *Helianthus strumosus*, with ovate leaves, whitish beneath, and the **Jerusalem artichoke**, *Helianthus tuberosus*, with leaves greenish beneath, and with tuberous rootstocks.

Sunflower

Helianthus mollis

Helianthus hirsutus

JERUSALEM ARTICHOKE
Helianthus tuberosus

Helianthus strumosus

Helianthus grosseserratus of low ground and swales is a tall species with more or less glaucous stems and alternate lanceolate leaves, whitish beneath, tapering to base of the blade. Other native sunflower species also occur in the Ozark region.

Importance The sunflowers range from fair to excellent forage for all kinds of livestock. The species associated with native bluestem ranges are all sensitive climax decreasers but usually are not abundant enough to be key indicators of range condition.

The sunflowers are generally fair wildlife plants. The flower heads and upper stems are nipped by deer during mid- to late-summer.

Ox-eye

Description Tall upright perennial herbs similar to sunflower in general appearance, with mostly rough stems; leaves simple, opposite, ovate to ovate-lanceolate, short-petioled or nearly sessile, distinctly dentate on the margins; flower heads showy with yellow rays surrounding a more or less conical receptacle, on terminal stalks, developing in early summer. In sunflower, *Helianthus*, the receptacle is mostly flat.

OX-EYE, *Heliopsis helianthoides*

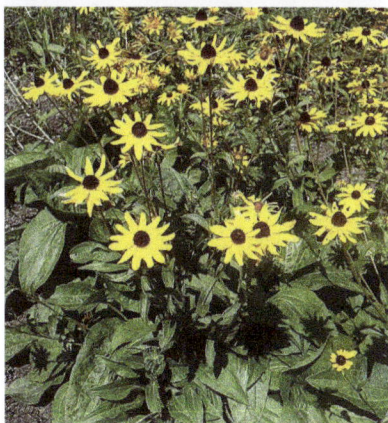

Distribution Widely scattered throughout the region, seldom abundant.
Habitat Prairies, glades, dry woods, and waste ground.
Importance Ox-eye ranges from poor to good forage for livestock.

Hawkweed

Description Upright perennial herbs with milky sap and pubescent stems and foliage, to 3 ft. (90 cm.) tall; leaves simple, mostly basal or alternate, oblong-lanceolate or spatulate, becoming bractlike upward, sessile, entire on the margin; flower heads with small yellow rays on short ascending branches of a terminal inflorescence, developing in spring and through summer.

Hieracium gronovii

Hieracium gronovii

Distribution Widely scattered throughout the region.
Habitat Glades, prairies, open woods, and waste places.

Two prevalent species are *Hieracium gronovii*, with relatively short pilose hairs less than ⅜ in. (9–10 mm.) long on the stem and leaves, and *Hieracium longipilum* with conspicuous white or rust-colored pubescence to ½ in. (10–12 mm.) long or more.

Importance Hawkweed is a fair forage plant for all kinds of livestock and is moderately palatable during spring and early summer. Deer eat seed heads and stem tips during summer and leaves during winter.

Lactuca spp.

Wild Lettuce

Description Mostly tall annual or biennial herbs with leafy stems and milky sap; leaves simple, alternate, sessile or petioled, lobed or deeply incised; flower heads with small yellow or bluish-purple rays, developing in summer.

WOODLAND LETTUCE
Lactuca floridana

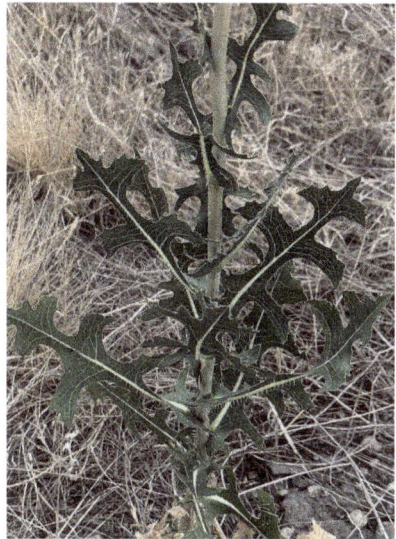

PRICKLY LETTUCE, *Lactuca serriola*

Distribution Widespread and sometimes common and abundant.
Habitat Fields, pastures, waste ground, open woods, bottomlands, and along streams; some species introduced.

Woodland lettuce, *Lactuca floridana*, is a tall conspicuous native species of woodlands with bluish flowers in a spreading panicle, and large, variously lobed or deltoid, pointed leaves.

Prickly lettuce, *Lactuca serriola* (synonym *Lactuca scariola*), introduced, has yellow flower heads and bristly, incised or lobed, clasping leaves and narrow panicles.

Two other species with yellow flowers are *Lactuca saligna* (introduced) and *Lactuca canadensis* (native) with softer spear-shaped or coarsely incised foliage, entire on the margin, not bristly; the first species with narrow unbranched flowered panicles, the latter more branching, with numerous flower heads.

Importance Wild lettuce is grazed heavily by deer. Wild lettuce is fair forage for cattle and good for sheep and goats, being eaten mostly in spring and early summer.

Bush-clover

Description Annual and perennial herbs with trailing or erect habit, some species to 4 ft. (1.2 m.) tall; leaves alternate trifoliate, stipulate but lacking the minute needlelike bracts below each leaflet that occur in *Desmodium*; flowers pink, purplish, creamy, or a combination, arranged in axillary clusters, loose terminal racemes, or dense heads, developing in late spring and summer; in some species less conspicuous apetalous flowers are also present; pod 1-seeded, flat and somewhat rounded, with pointed apex.

Lespedeza procumbens

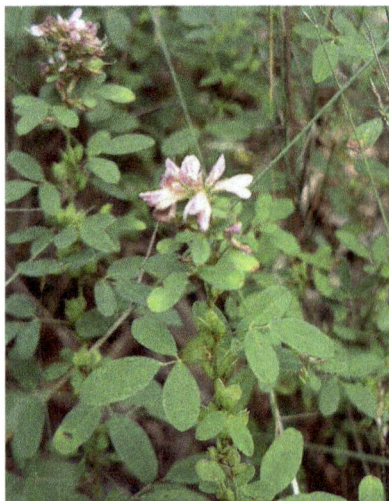

Lespedeza violacea

Distribution Widespread and sometimes common and abundant.
Habitat Dry open woods on cherty soils, also in glades, prairies, and, for introduced species, fields, roadsides, and waste ground.

Common native species are discussed below, followed by introductions.

Two species with low trailing habit are *Lespedeza procumbens* and *Lespedeza repens*, both with oval leaflets, the first species mostly pubescent, the second glabrous or nearly so.

Lespedeza violacea is a common Ozark species with slender erect stems, 20–25 in. (50–60 cm.) tall, branching from base and somewhat bushy, with oval leaflets to ¾ in. (20 mm.) wide.

A slender-stemmed lespedeza to 3 ft. (90 cm.) tall or more and one of the first species to flower in late spring is *Lespedeza virginica* with narrow, linear leaflets ⅛–¼ in. (3–6 mm.) wide.

Lespedeza hirta is a tall coarse heavy-stemmed species to 4 ft. (1.2 m.) tall or more, with pubescent foliage and oval or elliptic leaflets, and with flowers mostly whitish.

Lespedeza spp.

Bush-clover

Lespedeza capitata

Lespedeza hirta

Lespedeza capitata is distinguishable by dense crowded flower heads and elliptic-oblong leaflets with silvery or grayish pubescence, prevalent in glades, prairies, and upland woods.

Three introduced species are found in the region. **Chinese bush-clover**, *Lespedeza cuneata*, is a tall upright perennial forming dense stands in plantings, with grayish foliage and minute narrow leaflets. Two low annuals of fields and roadsides and spreading to waste ground are **Korean clover**, *Lespedeza stipulacea* (synonym *Kummerowia stipulacea*), and **Japanese clover**, *Lespedeza striata* (synonym *Kummerowia striata*), both with conspicuous stipules, the first species with petioled leaves, the second sessile or nearly so.

Importance The lespedezas provide excellent food for livestock and many kinds of wildlife. The seeds of most lespedezas are an important food for quail and ruffed grouse.

Lespedezas are readily eaten by deer, the native species primarily during summer, and Korean clover throughout the year.

The native lespedezas are especially high in crude protein and phosphorus content. They are considered decreasers on native ranges. These natives may succeed where many introduced legumes fail.

Japanese clover is planted in the south to raise the fertility level of fields and to increase the value of pasturage and hay. It can produce seed even when closely grazed because some stems hug the ground. The low-growing native plants are also somewhat resistant to grazing.

Blazing Star, Gayfeather

Description Upright perennial herbs with simple unbranched stems usually with a hard bulbous base; leaves simple, alternate, mostly linear or long and narrow, with entire margins; flowers small, roseate purple, sometimes white, arranged in dense elongate spikes or in compact clusters scattered along the stem, developing in summer.

Liatris pycnostachya

Distribution Widely scattered throughout the region.
Habitat Glades, prairies, open woods, bluffs.

Two common species with coarse stems are *Liatris pycnostachya* with numerous narrow leaves to 4 in. (10 cm.) long, and dense narrow flowering spikes as much as 10 in. (25 cm) long; and **tall gayfeather**, *Liatris aspera* with numerous, more or less globose heads in a raceme, spaced on the central stem.

Two other common species, also with distinct flower clusters spaced on the stem, are *Liatris cylindracea* and *Liatris squarrosa*.

Blazing Star, Gayfeather

TALL GAYFEATHER, *Liatris aspera*

Liatris cylindracea

Liatris squarrosa

Importance Gayfeather is nutritious and palatable to livestock during spring and early summer. It is considered a decreaser on native ranges.

The gayfeathers have a very colorful flower head and are often planted as ornamentals. The bulbs of some species have been used as food and medicine by the Native Americans.

Lobelia

Description Upright annuals or perennials 1–3 ft. (30–90 cm.) tall; stems simple or branching, with milky juice; leaves simple, alternate, sessile, serrate on the margins or only slightly so; flowers blue, red, or whitish, tubular, in terminal racemes or among the leaves, appearing in late spring and summer.

BLUE LOBELIA, *Lobelia siphilitica*

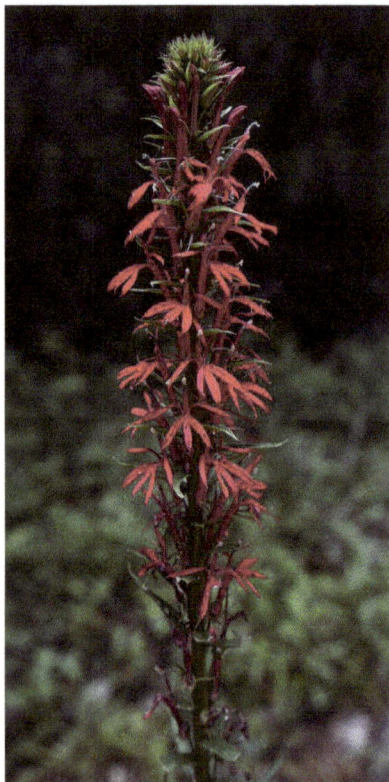

CARDINAL FLOWER
Lobelia cardinalis

Distribution Widely scattered throughout the region, seldom abundant.

Habitat Prairies, glades, wet or low ground, and moist habitats.

Blue lobelia, *Lobelia siphilitica*, common on low moist ground, is distinguished by coarse stems, usually branching, with ovate or oblong leaves and deep blue flowers in late summer.

Cardinal flower, *Lobelia cardinalis*, also of damp ground, has narrower lanceolate leaves and scarlet flowers, also in late summer.

Lobelia

Lobelia spicata

Lobelia spicata is identified by long, narrow, spikelike racemes of small pale-blue flowers, developing in late spring and generally earlier than the other species.

Indian tobacco, *Lobelia inflata*, is an annual species of open woods and fields, also low ground, with small bluish flowers and an inflated or subglobose calyx.

Importance Lobelia is a poor forage for livestock and may be poisonous. *Lobelia inflata* has been used as a medicine; overdoses may have the same effect as narcotic poison. Similar poisoning is also caused by *Lobelia siphilitica*.

INDIAN TOBACCO
Lobelia inflata

False Solomon's Seal, False Spikenard

Description Erect plants from a fleshy rhizome; with simple unbranched stems, 18–30 in. (50–75 cm.) tall; leaves simple, alternate, ovate-lanceolate, mostly sessile, with parallel veins and entire margins; flowers small, whitish, in erect terminal panicles above the leaves, appearing in spring; fruits globular, berrylike, reddish-translucent.

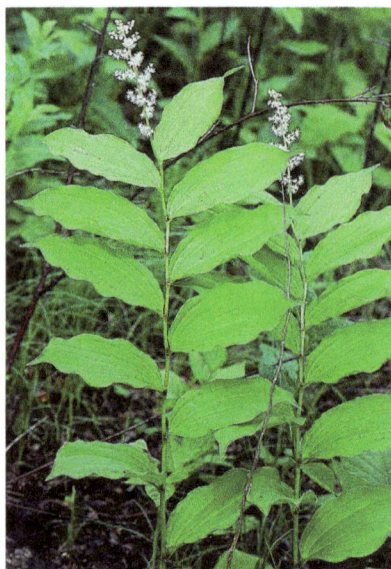

FALSE SOLOMON'S SEAL
Maianthemum racemosum

Synonym *Smilacina racemosa*
Distribution Widespread.
Habitat Wooded slopes, ravines, and bottomland.

Importance This plant provides occasional food for deer and turkeys.

Medicago lupulina

Black Medick

Description Low sprawling annual with angular stems to about 20 in. (50 cm.) long; leaves alternate, trifoliate, with finely serrate leaflets; flowers small, yellow, in elongate racemes, occurring throughout the season; pod circular, coiled.

BLACK MEDICK, *Medicago lupulina*

Distribution Widely scattered throughout the region, seldom abundant.

Habitat Fields, meadows, and waste ground; introduced from Eurasia and naturalized.

The cultivated **alfalfa**, *Medicago sativa*, is a taller upright plant with blue flowers, occasionally establishing in waste ground.

The low **hop clovers**, *Trifolium* spp., with yellow flowers and some-what resembling black medick, are identified by more or less spherical flower clusters and stems rounded in cross section.

Importance Black medick is highly nutritious and a good forage for livestock. It is probably eaten by deer, but no specific records are available for this area. It is grown to some extent for forage but generally has a low yield.

Sweet Clover

Description Tall annual or biennial, sweet-scented plants to 3–4 ft. (90–120 cm.); leaves alternate, trifoliate, minutely serrate on the margin, and with narrow pointed stipules at base; flowers white for *Melilotus albus*, and yellow for *Melilotus officinalis*, both with slender spikelike racemes, developing in spring and summer; pod small, short, with 1 or 2 seeds.

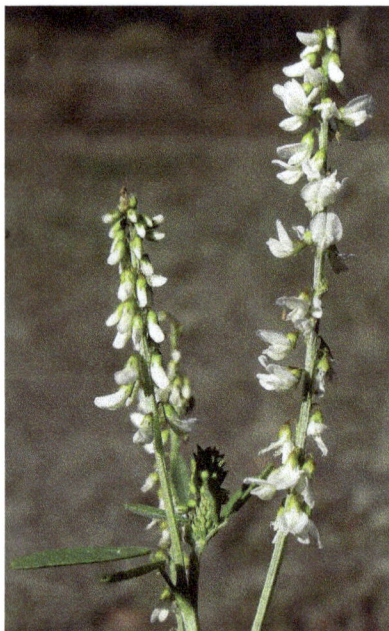

WHITE SWEET CLOVER
Melilotus albus

YELLOW SWEET CLOVER
Melilotus officinalis

Distribution Widespread and sometimes common and abundant.
Habitat Fields and waste ground; all introduced species.
Importance Before becoming rank, sweet clover is an excellent forage plant for livestock and is also eaten by deer. It is highly nutritious and commonly used as hay or pasturage.

Pheasants occasionally eat sweet clover seeds. Bees produce an excellent honey from sweet clover.

Nuttall's Mimosa

Description Low trailing perennial with tough prickly stems; leaves alternate, pinnately compound with numerous fine leaflets about 1/16 in. (1½ mm.) wide which fold inward when the plant is touched; flowers pinkish, numerous, in spherical clusters, developing in summer; pods elongate, prickly, several-seeded.

NUTTALL'S MIMOSA, *Mimosa nuttallii*

Synonym *Schrankia nuttallii*
Distribution Widely distributed.
Habitat Open woods, glades, prairies, waste ground; on dry soils.
Importance Sensitive brier is highly nutritious and readily eaten by all kinds of livestock in spring when the protein content is very high. It decreases under heavy grazing and is an important indicator of range condition.

Quail occasionally eat the seeds and turkeys eat the leaves.

Seeds from this plant contain a purgative and have been used in laxatives.

Monarda spp. **LAMIACEAE, MINT FAMILY**

Horsemint, Wild Bergamot, Beebalm

Description Upright annuals or perennials with strong minty odor; stems simple or branching, to 4 ft. (1½ m.) tall, leaves simple, opposite or whorled, ovate to linear-lanceolate, with serrate margins; flowers pale purple, lilac, or whitish, tubular, in terminal heads and axillary clusters, developing in spring and summer.

HORSEMINT, *Monarda fistulosa*

Monarda russeliana

LEMON MINT, *Monarda citriodora*

Distribution Widely scattered throughout the region.
Habitat Glades, open woods, fields, and low ground.

Two tall branching perennials are horsemint, *Monarda fistulosa*, with ovate petioled leaves and pinkish flowers, and *Monarda russeliana*, with mostly simple stems and nearly sessile leaves (but absent from Missouri).

Lemon mint, *Monarda citriodora*, an annual with narrow whorled leaves, is less widespread than the above species (becoming more common in the southwestern USA), and occurs most often in limestone glades and barrens.

Importance *Monarda* is poor to fair forage for cattle and fair for sheep and goats. Intense summer use may occur where deer numbers are high.

Oxalis spp.

Wood Sorrel

Description Low annuals or perennials with sour-tasting foliage; leaves trifoliate, on stem or from base of plant, the leaflets notched at apex; flowers yellow or purplish with 5 similar petals, appearing in spring and summer.

YELLOW WOOD SORREL
Oxalis stricta

CREEPING WOOD SORREL
Oxalis corniculata

VIOLET WOOD SORREL
Oxalis violacea

Distribution Widely distributed.
Habitat Glades, open woods, fields, waste ground, and stream bottoms.

Yellow wood sorrel, *Oxalis stricta*, and creeping wood sorrel, *Oxalis corniculata*, are yellow-flowered species with leaves on stem. The latter species has spreading horizontal stems, rooting at the nodes.

Violet wood sorrel, *Oxalis violacea*, is identified by purplish flowers, the leaves all arising at base of plant from a scaly underground bulb.

Importance Sorrel is fair forage for cattle and horses and good for sheep and goats. However, the plants usually do not furnish sufficient volume of forage to be of much importance. Sorrel is eaten by wild turkeys for a short period in spring and occasionally by prairie chickens and quail.

Golden Ragwort, Squaw-weed

Description Upright perennials; leaves dissimilar, those at ground level petioled, oval, cordate, or obovate, serrate on the margin, those on the stem dissected; flower heads with yellow rays on ascending branches from upper stem, developing in spring and summer.

GOLDEN RAGWORT, *Packera aurea*

Packera obovata

Distribution Widely distributed.
Habitat Prairies, meadows, low woods, streambanks, and waste ground.

Two common species are **golden ragwort**, *Packera aurea* (synonym *Senecio aureus*), identified by mostly cordate basal leaves with scalloped margins, and squaw-weed, *Packera obovata* (synonym *Senecio obovatus*), with obovate basal leaves gradually tapering to petiole.

Prairie ragwort, *Packera plattensis* (synonym *Senecio plattensis*), has ovate basal leaves tapering abruptly to the petiole.
Importance Deer eat the basal leaves of ragwort during the winter. The Ozark species are poor forage for livestock and will invade depleted ranges.

PRAIRIE RAGWORT
Packera plattensis

Parthenium integrifolium

American Feverfew

Description Erect strong-rooted perennials with leafy stems to about 3 ft. (90 cm.) tall; leaves simple, alternate, ovate to lanceolate, with long tapering petioles at base of plant, becoming sessile and decreasing in size on upper stem, serrate on the margin; flower heads small, whitish, in a branching terminal inflorescence, developing in spring and summer.

AMERICAN FEVERFEW, *Parthenium integrifolium*

Distribution Widely scattered throughout the region, seldom abundant.

Habitat Glades, prairies, and dry woods.

Importance *Parthenium* provides fair to good forage for cattle, horses, sheep, and goats.

Beardtongue, Penstemon

Description Upright perennials, stems solitary or clumped; leaves simple, opposite, mostly sessile, serrate or entire on the margin; flowers whitish or blue, showy, tubular, in terminal panicles, developing in spring and summer; fruit a capsule with numerous minute seeds.

PALE BEARDTONGUE
Penstemon pallidus

FOXGLOVE BEARDTONGUE
Penstemon digitalis

Distribution Widespread, or more local for *Penstemon cobaea*.
Habitat Glades, prairies, open woods, swales, low ground, and alluvium.

Two common species are **pale beardtongue**, *Penstemon pallidus*, with pubescent foliage, and **foxglove beardtongue**, *Penstemon digitalis*, a taller glabrous plant to 3–4 ft. (90–120 cm.) or more, both with whitish flowers.

Arkansas beardtongue, *Penstemon arkansanus*, also with white flowers, is found on rocky or sandy soils in open woodlands.
Importance Penstemons are good to forage plants for cattle and excellent for sheep and goats, especially during the early spring.

ARKANSAS BEARDTONGUE
Penstemon arkansanus

A number of penstemons are planted as ornamentals because of their showy flower heads.

Knotweed, Smartweed

Description Low or spreading to tall erect branching annuals and perennials of the buckwheat family, with sheathing stipules (ocreae); leaves simple, alternate, of various shapes and sizes, entire on the margin; flowers small, greenish white to pink or roseate, in narrow spikelike racemes or in axillary clusters, appearing throughout the season; fruit an achene, brownish or black, shiny, flat, and circular or triangular.

Persicaria lapathifolia

Persicaria punctata

Distribution Widespread and sometimes common and abundant.
Habitat Fields, waste areas, sloughs, wet ground, rich woods, streambanks and alluvium; several species naturalized.

Common species include ***Persicaria pensylvanica*** (synonym *Polygonum pensylvanicum*) and ***Persicaria lapathifolia*** (synonym *Polygonum lapathifolium*), both tall annuals to 4 ft. (1.2 m.) or more, with mostly lanceolate leaves. The first species has pink erect spikes; the latter drooping, somewhat greenish spikes.

Persicaria punctata (synonym *Polygonum punctatum*) and ***Persicaria maculosa*** (synonym *Poly-gonum persicaria*) are generally shorter plants with lanceolate leaves and fringed or lacerated stipular sheaths, the first species with greenish-white flowers, and the latter with generally pink flowers.

The common **knotweed, *Polygonum aviculare***, is a low spreading annual with small bluish-green leaves and with flowers more or less hidden in the axils of the leaves, common in pastures, waste ground, and lawns.

Persicaria virginiana (synonym *Polygonum virginianum*), typical of rich woods, is rhizomatous with broadly elliptic pointed leaves, and slender, sparsely flowered spikes.

Knotweed, Smartweed

Persicaria maculosa

Persicaria virginiana

KNOTWEED, *Polygonum aviculare*

SLENDER KNOTWEED
Polygonum tenue

Slender knotweed, *Polygonum tenue,* of dry rocky sites, is identified by short slender stems about 15 in. (40 cm.) tall, and narrow leaves.

Importance Smartweed seeds are eaten by ducks, geese, pheasants, mourning doves, quail, and prairie chickens. Ruffed grouse, deer,

turkeys, and raccoons also utilize plants of this genus.

These plants are of little value to livestock, and some are suspected of being poisonous to horses, sheep, and hogs, but proof is lacking. They may also cause dermatitis in some people.

Phlox spp.

Phlox

Description Low perennials, somewhat tufted, with erect or decumbent stems; leaves simple, opposite, ovate-lanceolate to narrow-linear, entire on the margin; flowers blue, roseate, or white in loose more or less terminal clusters, appearing in spring.

DOWNY PHLOX, *Phlox pilosa*

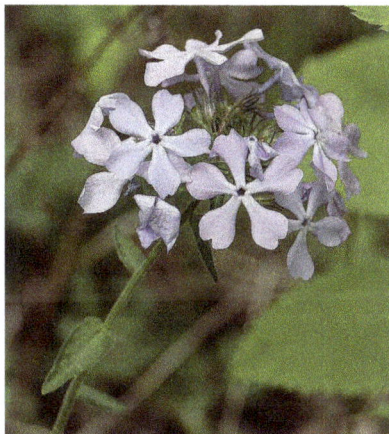

WILD BLUE PHLOX, *Phlox divaricata*

Distribution Widely distributed.
Habitat Dry upland woods, prairies, glades, or shaded slopes and ravines.

Downy phlox, *Phlox pilosa*, with linear pointed leaves and pink or roseate flowers, is common in dry woods, prairies, and glades.

Wild blue phlox, *Phlox divaricata*, is a most common species identified by wider ovate-lanceolate leaves, spreading stems, and blue flowers, more generally on shaded sites and low ground.

Importance Phlox is good forage for sheep and goats, but cattle eat it only very early in the spring when better forage is scarce. Deer eat the overwintering green leaves. Ruffed grouse and prairie chickens occasionally eat the seed-filled capsules.

Ground Cherry

Description Low or tall branching annuals or perennials, to about 3 ft. (90 cm.) or more, the perennials usually rhizomatous; leaves simple, alternate, ovate to broadly lanceolate, nearly entire or wavy margined to coarsely dentate; flowers yellow with brownish or purplish center, developing in summer; berry yellow, reddish, or purple, many-seeded, enclosed by inflated, saclike calyx.

GROUND CHERRY
Physalis pubescens

Physalis longifolia

Physalis virginiana

Distribution Widely scattered throughout the region, seldom abundant.

Habitat Glades, prairies, open woods, waste ground, bottomlands, alluvium.

The annual **ground cherry**, *Physalis pubescens*, has low villous stems and rounded or cordate leaves mostly entire on the margins. Rhizomatous perennials include *Physalis longifolia*, mostly glabrous, with entire or wavy-margined lanceolate leaves, and two pubescent or hairy-stemmed plants: *Physalis heterophylla* with rounded or cordate leaves, and *Physalis virginiana* with narrow leaves tapering at base.

Importance Ground cherry provides small amounts of food for quail, pheasant, deer, prairie chickens, and turkeys. It is poor to worthless as forage for livestock. Livestock have been poisoned, but seldom fatally, by eating large amounts of tops and green berries.

Phytolacca americana

Pokeweed

Description Coarse branching perennial, strong smelling, with thick glabrous reddish-purple stems to 6–7 ft. (2 m.) tall or more; leaves large, ovate, petioled, with entire margins; flowers whitish, in elongate racemes, developing in late spring and through the summer; fruit a dark purplish berry, numerous, in elongate racemes.

POKEWEED, *Phytolacca americana*

Distribution Widespread and sometimes common and abundant.
Habitat Fields, waste places, disturbed ground, and clearings.
Importance The small, brightly colored berries are eaten by pheasants mourning doves, ruffed grouse, raccoons, and occasionally by prairie chickens and quail. Deer occasionally eat the plants, especially during spring and early summer.

Pokeweed is grazed by cattle early in summer but is only poor to fair forage.

Young shoots are cooked and eaten as greens by humans, but the roots, older stems, and seeds reportedly have poisonous properties.

The juice from the ripe berries was formerly used as an ink substitute.

Buckhorn, Plantain, Ribgrass

Description Low annuals, some species biennial or perennial, with basal rosette of simple parallel-veined leaves, entire on the margin; flower stalks solitary or several, from ground level, with narrow terminal spikes of inconspicuous flowers, developing in spring and through the summer.

BRACTED PLANTAIN
Plantago aristata

Plantago rugelii

Distribution Widespread and sometimes common and abundant.
Habitat Dry fields, pastures, glades, and waste ground; on thin soils; *Plantago lanceolata* introduced from Europe.

The common **bracted plantain**, *Plantago aristata*, of sterile habitats has linear ascending leaves, sometimes only 1/16 in. (1½ mm.) wide, mostly pubescent, and stiff spikes with conspicuous attenuate bracts subtending the flowers.

Buckhorn, **ribgrass**, or **English plantain**, *Plantago lanceolata*, is a coarser plant, identified by mostly lanceolate leaves to about 1½ in. (4 cm.) wide with long tapering petioles and inconspicuous oval bracts on the flowering spike.

Plantago rugelii is a common perennial with broadly ovate smooth leaves, and reddish-purple petioles.
Importance Plantain is worthless to poor as a forage plant for livestock. It invades abused native ranges and low-vigor tame pastures. It is usually eaten only when other more palatable forage is not available.

Occasionally the seed heads are eaten by deer, and the seeds and leaves by prairie chickens.

Christmas Fern

Description Fronds spreading or arching, elongate, 12–24 in. (30–60 cm.) long, dark-green, firm-textured; base of individual pinnae (frond divisions) with characteristic projection on upper margin (toward the apex of frond); central axis of frond somewhat scaly or chaffy.

CHRISTMAS FERN, *Polystichum acrostichoides*

Distribution Widely scattered throughout the region, seldom abundant.

Habitat Shaded slopes and ravines; on generally well-drained soils.

Importance Deer eat small amounts of Christmas fern during winter; wild turkeys feed upon the fronds during summer and fall. Ruffed grouse eat this fern during winter and spring.

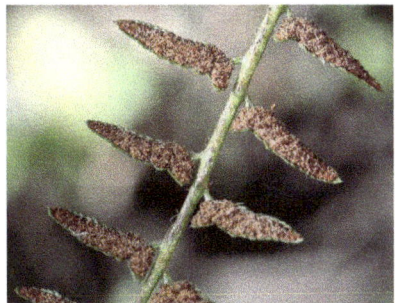

Underside of fertile frond

Cinquefoil

Description Perennial herbs, mostly pubescent, with stems erect or trailing and rooting at the tips; leaves alternate, compound, with 3–7 or more serrate or coarsely toothed leaflets, and conspicuous stipules at base of petioles; flowers yellow, with 5 equal petals, appearing in spring and summer; fruit dry, with numerous achenes.

OLDFIELD CINQUEFOIL
Potentilla simplex

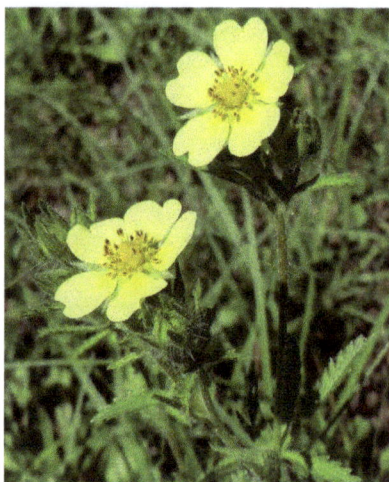

SULPHUR CINQUEFOIL
Potentilla recta

Distribution Widely scattered, seldom abundant.

Habitat Glades, prairies, dry woods, fields, and waste ground.

Two common species with palmately compound leaves are the native **oldfield cinquefoil**, *Potentilla simplex*, and the introduced **sulphur cinquefoil**, *Potentilla recta*, the first species with trailing stems, the latter with erect stems.

Tall cinquefoil, *Potentilla arguta* (synonym *Drymocallis arguta*), a less common species in the Ozarks, has coarse hairy stems to 3 ft. (90 cm.) tall or more, and pinnate leaves with 7 or more leaflets.

Importance Cinquefoils are grazed

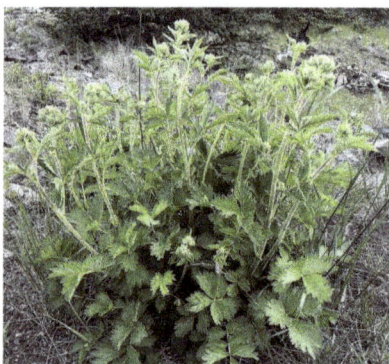

TALL CINQUEFOIL
Potentilla arguta

by deer during winter and early spring. They provide fair forage for sheep. The leaves and stems are high in tannic acid content, which may account for their low palatability.

Prunella vulgaris
Self-heal, Heal-all

Description Perennial herb of the mint family but not aromatic, somewhat tufted, with mostly simple stems to about 20 in. (50 cm.) tall; leaves simple, opposite, ovate-lanceolate with wavy or entire edges; flowers bluish, in spicate clusters, appearing in spring and summer.

SELF-HEAL, *Prunella vulgaris*

Distribution Widely distributed. **Habitat** Fields, waste areas, low ground, and streambanks; mostly on damp soils.

Importance This plant has little value for wildlife and is eaten by livestock only when more palatable forage is not available.

Cudweed, Everlasting

Description Upright whitish-woolly perennials, somewhat tufted, the stems simple or branching, to 20 in. (50 cm.) tall; leaves simple, alternate, sessile, linear-lanceolate to obovate or spatulate; flower heads small, woolly, in terminal clusters, developing in spring and summer.

Pseudognaphalium obtusifolium

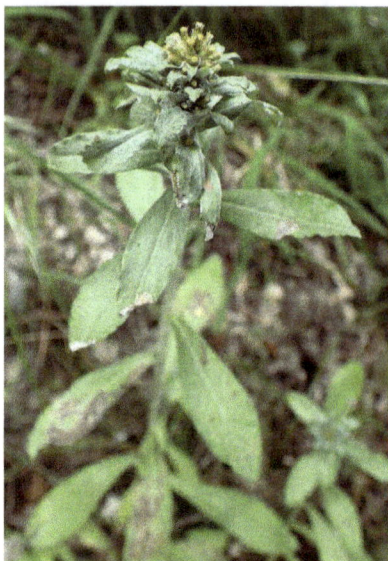

Gamochaeta purpurea

Distribution Widely distributed.
Habitat Glades, prairies, open woods, and waste areas.

Two similar species are prevalent (these formerly treated as *Gnaphalium*): **Pseudognaphalium obtusifolium** (synonym *Gnaphalium obtusifolium*), with linear leaves and branching stems, and **Gamochaeta** *purpurea* (synonym *Gnaphalium purpureum*), with leaves broader, obovate or spatulate, and stems mostly unbranched.
Importance Cudweed is poor to fair forage for livestock, being more palatable to sheep and goats than cattle.

former *Psoralea* spp.

Prairie Turnip, Sampson's Snakeroot, Scurf-pea

Description Deep-rooted perennials with erect stems to about 3 ft. (90 cm.) tall; leaves alternate, compound, with 3–5 linear to oblong leaflets, entire on the margin; flowers pale-blue or purple in loose racemes or spicate heads, developing in late spring and summer; fruit minute, 1-seeded.

SAMPSON'S SNAKEROOT, *Orbexilum pedunculatum*

Distribution Widespread, some species with limited range.
Habitat Prairies, glades, open woods; on acidic or alkaline soils.

Sampson's snakeroot, *Orbexilum pedunculatum* (synonym *Psoralea psoralioides* var. *eglandulosa*) occurs throughout the range, generally on cherty or sandy soils, and usually has 3 narrow leaflets 2½ in. (6–7 cm.) long, and flowers on conspicuously long peduncles.

Two other species are **scurf-pea**, *Psoralidium tenuiflorum* (synonym *Psoralea tenuiflora*), mainly in the western part of the region, and **prairie turnip**, *Pediomelum esculentum* (synonym *Psoralea esculenta*), with palmately compound leaves, which is sporadic. *Psoralidium tenuiflorum* is somewhat bushy, with scurfy-pubescent stems about 3 ft. (90 cm.) tall, the flowers from ends of leafy branches. *Pediomelum esculentum* is shorter, about 20 in. (50 cm.) tall, with villous hairs and an enlarged taproot.

former *Psoralea* spp.

Prairie Turnip, Sampson's Snakeroot, Scurf-pea

PRAIRIE TURNIP, *Pediomelum esculentum*

Importance Scurf-pea is usually not relished by livestock but is eaten occasionally in the early stages of growth when other more palatable forage is not present. It cannot withstand much grazing and soon disappears from overgrazed range. *Psoralidium tenuiflorum* is readily eaten when cured in hay. The foliage is suspected of being poisonous to livestock if eaten in large amounts, but actual records of poisoning are rare.

The large taproots of *Pediomelum esculentum* are edible and were an important food for the Native Americans who reportedly ate it

SCURF-PEA, *Psoralidium tenuiflorum*

both raw and cooked. Native Americans also dried the roots and made a powdery meal, which kept well for a long time when stored. The roots reportedly contain about 70 percent starch and 5 percent sugar and are quite nourishing.

Pteridium aquilinum

Bracken Fern

Description Erect coarse plants spreading by rhizomes, to 3–4 ft. (90–120 cm.) tall, dividing at top into broadly triangular, flat-topped, shiny fronds with numerous segments with indented edges.

BRACKEN FERN, *Pteridium aquilinum*

Distribution Widely distributed.
Habitat Pine and oak woodland and disturbed areas; on acid soils; increasing after fire.
Importance In dry seasons when other forage is scarce, livestock will graze the fronds. When eaten in hay over a long period, bracken fern may be fatal. It may cause high temperature, hard breathing, nasal bleeding, fever, rash, and hemorrhage of various organs. It may also be a photosensitization agent to cattle, causing blisters, blindness, sore mouths, and malnutrition after exposure to sunlight. When the range is in poor condition and the ferns are present, bulky supplemental feed should be provided or cattle should be removed from the range.

Bracken fern usually dries in late summer and is a fire hazard where it occurs in dense stands.

Mountain Mint

Description Upright perennial herb with minty odor and square stems to 3 ft. (90 cm.) tall; leaves simple, opposite, linear-oblong, with entire margins; flowers small, tubular, whitish, in more or less dense terminal clusters, developing in summer.

MOUNTAIN MINT, *Pycnanthemum tenuifolium*

Distribution Widespread and sometimes common and abundant.
Habitat Dry upland woods, prairies, glades, and low ground.

Stenaria nigricans (synonym *Houstonia nigricans*) of similar habitats, which might be mistaken for mountain mint, is generally a smaller plant, lacking the minty odor and having regular flowers, the 5 petals all equal.
Importance Mountain mint is eaten by livestock only in the very early spring or late fall when more palatable forage is scarce or dry.

Ranunculus spp.

Buttercup

Description Upright annual or perennial herbs, somewhat tufted, with leafy bases, to 3 ft. (90 cm.) tall, or creeping with horizontal stems. Two groups occur, those with dissected leaves throughout the plant, on the stem and from the base, and those with dissected leaves on stem and generally rounded or reniform leaves with scalloped margins at base of plant; flowers with 5 yellow petals, all similar, developing in spring.

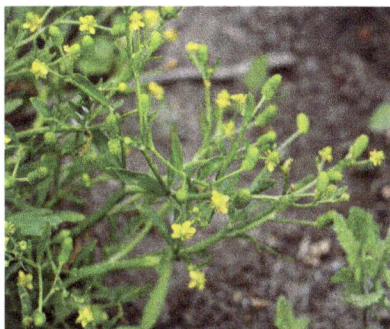

CURSED CROWFOOT
Ranunculus sceleratus
ABOVE, RIGHT

Distribution Widespread, or some species with limited range.
Habitat Woods, streambanks, low ground; on moist or dry soils.

Cursed crowfoot, *Ranunculus sceleratus*, a wetland plant uncommon in the Ozarks, has dissected leaves, glabrous, somewhat fleshy stems, and small flowers with petals not exceeding ¼ in. (5–6 mm.).

A buttercup, also of damp or dry ground with much larger flowers, the petals ½ in. (12 mm.) long or more, and with stem and basal leaves dissected is *Ranunculus hispidus* (synonym *Ranunculus septentrionalis*).

Early buttercup, *Ranunculus fascicularis*, usually in dry places, and also with large flowers and dissected leaves, is characterized by short fleshy tuberlike roots.

Two buttercups with stem and basal leaves dissimilar, the upper stem leaves mostly dissected, the lower basal ones rounded and scalloped on the edges, are *Ranunculus abortivus* with glabrous leaves and stems, and *Ranunculus micranthus* with pubescent leaves and stem, both species with minute flowers. *Ranunculus abortivus* usually occurs on lower and moister ground compared to *R. micranthus*.

Buttercup

Ranunculus hispidus

EARLY BUTTERCUP
Ranunculus fascicularis

Ranunculus abortivus

Ranunculus micranthus

Importance *Ranunculus abortivus* and *R. hispidus* comprise an important part of the April through June diet of wild turkeys. *Ranunculus fascicularis* provides food for prairie chickens during these same months. Buttercups are important ruffed grouse foods during June. Deer eat the basal leaves, which remain green overwinter.

Buttercup is worthless to poor forage for livestock, being taken only when more palatable plants are absent or scarce. Eating these leaves before they are dried may inflame the intestinal tract. Cattle are most commonly affected; the milk may be bitter or reddish.

Cursed crowfoot and perhaps some of the other buttercups contain a substance strong enough to cause skin blisters on susceptible people.

Coneflower, Black-eyed Susan

Description Upright perennial and biennial herbs with leafy stems, occasionally tufted or with rosette of basal leaves; leaves simple, alternate, entire or dentate on the margin, or deeply lobed; flower heads terminal, several, with showy yellow rays surrounding a dark conical center, developing in spring and summer.

BLACK-EYED SUSAN, *Rudbeckia hirta*

Distribution Widely distributed.
Habitat Gades, prairies, open woods, waste areas, to low ground and streambanks.

A most common species of fields and waste ground is the **black-eyed Susan**, *Rudbeckia hirta*, with hirsute stems and foliage, the leaves sessile, narrowly ovate to lanceolate, tapering toward base.

A species of limestone glades and prairies, *Rudbeckia missouriensis*, has villous linear-lanceolate leaves.

Two common species with lobed leaves are *Rudbeckia laciniata* with smooth stems to 6–7 ft. (2 m.) tall or more, with deeply lobed glabrous leaves and pale yellow drooping rays, found in low or moist woods, and *Rudbeckia triloba* with mostly pubescent foliage, the upper leaves uncleft, which occurs in both open woods and on low ground.

Coneflower, Black-eyed Susan

Rudbeckia missouriensis

Rudbeckia laciniata

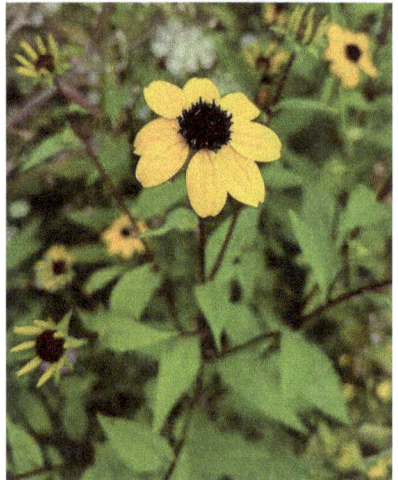

Rudbeckia triloba

The yellow-rayed **coneflower**, *Echinacea paradoxa*, is easily separated from *Rudbeckia* by its linear-lanceolate leaves with long narrow tapering leaf stalks.

Importance Basal leaves of these plants remain green overwinter and provide an important winter food for deer.

Coneflowers provide only poor to fair forage for livestock and are often considered range pests.

Ruellia, Wild Petunia

Description Perennial herbs; leaves simple, opposite, oval to oblong-lanceolate, mostly sessile or with short petiole, entire on the margin; flowers large, bluish, petunialike, in the axils of the leaves, developing in spring and summer.

Ruellia humilis

Ruellia pedunculata

Ruellia pedunculata

Distribution Widely distributed.
Habitat Glades, prairies, and open woods, or ravines, low woods, and stream bottoms.

A common species of glades and open woods is **Ruellia humilis**, with branching spreading stems about 15–20 in. (35–50 cm.) tall, conspicuously pubescent foliage, and flowers without stalks.

Ruellia pedunculata differs with sparsely pubescent stems and flowers on extended stalks.

Ruellia strepens of ravines and rich woods is identified by tall, mostly glabrous stems (or stems sparsely hairy along two opposite lines) to about 3 ft. (90 cm.), and stalked flowers.

Importance These plants have little food value for game or livestock.

Rumex spp.

POLYGONACEAE, BUCKWHEAT FAMILY

Dock, Sorrel

Description Small, slight, or coarse, thickstemmed perennials of the buckwheat family with sheathing stipules (ocreae), some species with rosette of basal leaves; leaves simple, alternate, of various shapes and sizes; flowers greenish or reddish brown in crowded spikelike racemes, appearing throughout the season; achene or fruit dry, 3-angled.

SOUR DOCK, *Rumex crispus*

Rumex obtusifolius

Distribution Widespread and sometimes common and abundant.
Habitat Fields, pastures, waste ground, sloughs, and streambanks.

Two common species with large more or less crinkled leaves are **sour dock**, *Rumex crispus*, with narrow elliptic-lanceolate basal leaves, and *Rumex obtusifolius*, with large cordate leaves, blunt at the apex, and long-petioled.

Rumex altissimus, also coarse and with large leaves, is identified by flat, somewhat fleshy, narrow-oblong leaves lacking wavy edges.

Red sorrel, *Rumex acetosella*, is a low tufted plant with hastate or arrow-shaped leaves and reddish flower stalks, common on thin acid soils, and spreading by slender rhizomes.

178 | WILDFLOWERS

Dock, Sorrel

Rumex altissimus

RED SORREL, *Rumex acetosella*
TOP, ABOVE

Importance In a Missouri study *Rumex acetosella* was the most abundant July food for turkeys. These plants provide food for deer during the summer, and small amounts are eaten by prairie chickens and pheasants.

It is a poor forage plant for livestock, being taken only in early spring when other feed is scarce.

Rumex acetosella and R. *crispus* may cause dermatitis to some people.

Bloodroot

Description Low perennial with thick horizontal rootstock having reddish sap and a single leaf rising from ground level; leaf rounded, deeply lobed, with light-colored lower surface; flowers solitary, from base of plant with showy white petals developing in early spring.

BLOODROOT, *Sanguinaria canadensis*

Distribution Widespread and sometimes common.
Habitat Wooded slopes and ravines.
Importance Bloodroot is worthless to poor as a forage plant for livestock, being taken only when more palatable forage is absent or scarce. The rhizome is poisonous but is seldom eaten.

Silphium spp.

Compass Plant, Rosinweed, Carpenter's Square

Description Tall erect perennials with coarse stems and leaves; leaves simple, alternate or opposite or from base of plant, toothed or entire on the margin; flower heads with yellow rays, on thick peduncles, appearing in summer.

CUP PLANT, *Silphium perfoliatum*

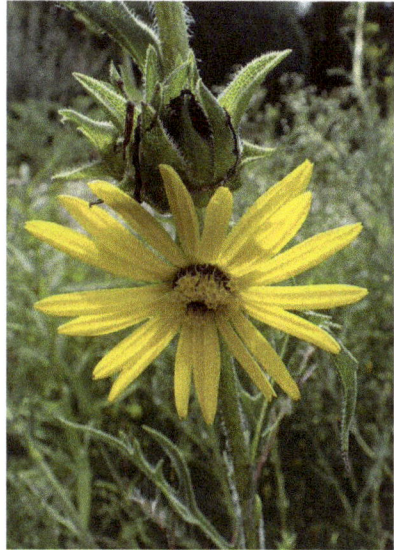

COMPASS PLANT
Silphium laciniatum

Distribution Widely distributed. **Habitat** On various sites including glades, prairies, rocky woods, low ground, and streambanks.

Two common species are **cup plant** or **carpenter's square**, *Silphium perfoliatum*, of low woods and damp ground, identified by thick, 4-angled stems and opposite leaves joined at base and surrounding the stem, and **compass plant**, *Silphium laciniatum*, of glades and prairies, with alternate deeply incised leaves, and harsh rough stems.

The easily recognized **prairie-dock**, *Silphium terebinthinaceum*, of prairies and limestone barrens, has large cordate basal leaves and generally leafless stems.

Two other species are **rosinweed**, *Silphium integrifolium*, with opposite ovate sessile leaves, mostly entire on the margins, and **starry rosinweed**, *Silphium asteriscus*, with alternate ovate to lanceolate leaves, also sessile, and usually dentate on the margin.

Compass Plant, Rosinweed, Carpenter's Square

ROSIN-WEED
Silphium integrifolium

PRAIRIE-DOCK
Silphium terebinthinaceum

STARRY ROSINWEED
Silphium asteriscus

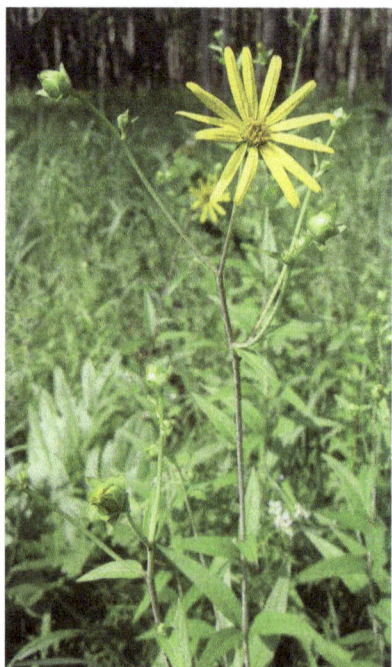

Importance These plants range from poor to good forage. Compass plant, the most important forage plant in the group, is readily eaten by all kinds of livestock, particularly when the plants are young. Compass plant is also a good indicator of range condition and disappears with overgrazing. The others mentioned above will increase with overgrazing.

Buffalo Bur, Horse Nettle, Nightshade

Description Low spreading or erect branching annuals or perennials with smooth or prickly stems; leaves simple, alternate, ovate or lobed, dentate or entire on the margins; flowers white, purplish, or yellow with 5 equal petals and yellow stamens, developing in spring and summer; fruit smooth or prickly.

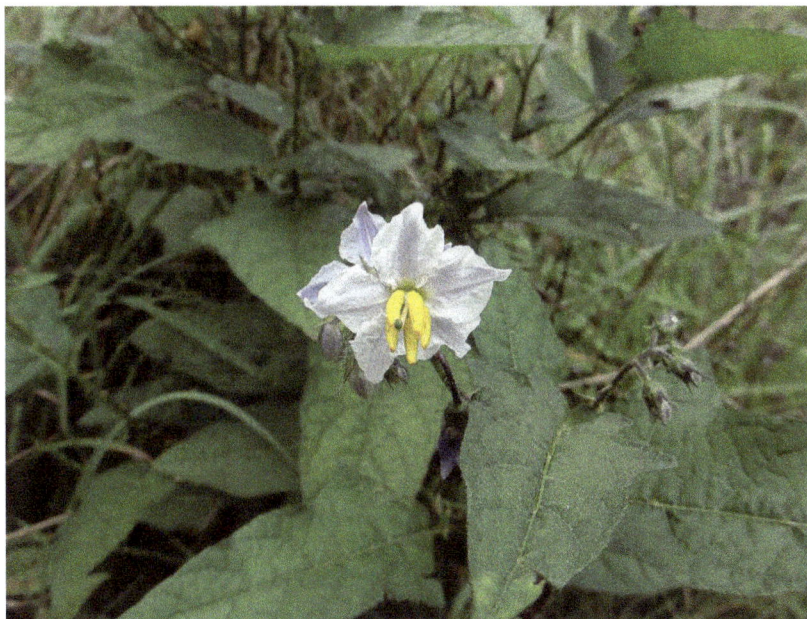

HORSE NETTLE, *Solanum carolinense*

Distribution Widespread and sometimes common and abundant.
Habitat Fields, pastures, waste ground, also open woods, prairies, and low ground.

The common **horse nettle**, also called **bull nettle**, *Solanum carolinense*, is a perennial with prickly stems, purplish-white flowers, and smooth yellow berries.

The annual **black nightshade**, *Solanum americanum*, has smooth foliage, small white flowers, and smooth dark fruits.

The less common **buffalo bur**, *Solanum rostratum*, is a conspicuously prickly annual with yellow flowers, deeply lobed or divided leaves, and spiny fruits.

Buffalo Bur, Horse Nettle, Nightshade

BUFFALO BUR, *Solanum rostratum*

BLACK NIGHTSHADE
Solanum americanum

Importance The fruit and seeds of horse nettle and black nightshade are used as food by prairie chickens, and to some extent by ruffed grouse, quail, deer, turkeys, pheasants, and raccoons. The plants are eaten by livestock only when other forage is scarce. Black nightshade fruits and foliage are poisonous to livestock, especially during July and August. Toxicity decreases as the fruits ripen.

Buffalo bur is an especially aggressive drought-resistant plant and produces a fruit with sharp spines that can injure the animals eating it.

The annuals can be controlled by mowing before seed is matured.

Chemical sprays and dense stands of climax grasses will limit the abundance of the perennial species of Solanum.

The ripe berries of black nightshade are sometimes cooked for preserves or jams. Cooking apparently destroys the toxic material present.

Solidago spp.

Goldenrod

Description Low to tall coarse perennials with mostly simple stems; leaves simple, alternate, of various shapes and sizes, mostly sessile, or short petioled toward base of plant, serrate or entire on the margin; flowers small, yellow, in open branching or narrow somewhat compact panicles, developing in summer and fall.

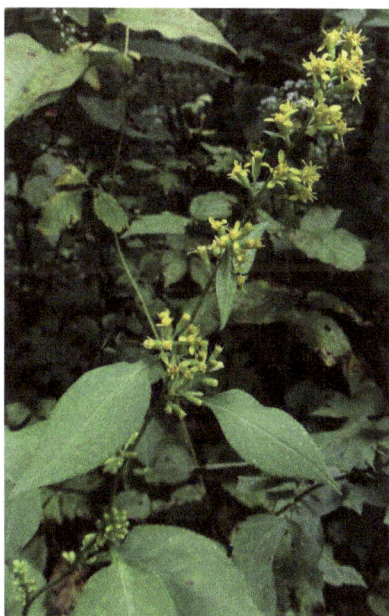

Distribution Widespread and sometimes common and abundant.
Habitat Glades, prairies, open woodland and bluffs, ravines, also low rich woods and deep soils.

A common species with broadly ovate leaves is **broadleaf** or **zigzag goldenrod**, *Solidago flexicaulis*, of rich woods, with oval leaves and short winged petiole.

Hairy goldenrod, *Solidago hispida,* of drier woods on acid soils, is distinguished by pubescent foliage and stems, and narrow obovate basal leaves.

Solidago arguta of dry woodlands has glabrous foliage and ovate basal leaves.

Solidago radula, a common species of dry woods and bluffs, has rough elliptic leaves, smaller upward, with few teeth, and a panicle

BROADLEAF GOLDENROD
Solidago flexicaulis
ABOVE LEFT, ABOVE

with short ascending branches.

Still-leaved goldenrod, *Solidago rigida*, is a glade and prairie species with harsh oval to elliptic sessile leaves and a flat-topped inflorescence.

A characteristic species of old fields, open ground, and glades is **gray goldenrod**, *Solidago nemoralis*, with narrow lanceolate to spatulate leaves and spreading arching inflorescence. Other species are also present in the Ozarks.

Solidago spp.

ASTERACEAE, ASTER FAMILY

Goldenrod

HAIRY GOLDENROD
Solidago hispida

Solidago arguta

Solidago radula

Importance The basal leaves of goldenrod are commonly eaten during late fall or winter by deer, and to a lesser extent by ruffed grouse and turkeys; the latter also eat the young flower buds in summer and autumn. These plants make poor to fair forage for livestock and are often abundant on overgrazed lands.

Solidago spp.
Goldenrod

STIFF-LEAVED GOLDENROD, *Solidago rigida*

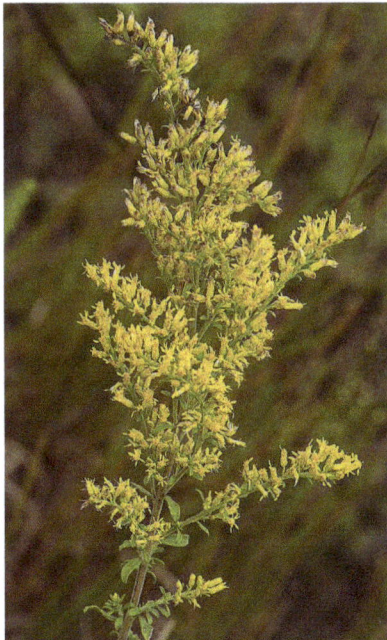

GRAY GOLDENROD
Solidago nemoralis
ABOVE, RIGHT

Diamondflowers

Description Upright perennial herb with square stems, usually several from a hard base, about 20 in. (50 cm.) tall or more; leaves simple, opposite, linear, single-veined, entire on the margin, blackening upon drying; flowers small, tubular, pinkish white, in short-branching clusters from the middle and upper stem, in late spring through summer.

DIAMONDFLOWERS, *Stenaria nigricans*

Synonyms *Hedyotis nigricans, Houstonia nigricans*
Distribution Widespread and sometimes common and abundant.
Habitat Limestone glades, bluffs, and woods openings; on dry soils.
Importance Prairie chickens occasionally eat the seeds of *Stenaria*. It is poor to fair forage for livestock, being eaten occasionally in spring and early summer.

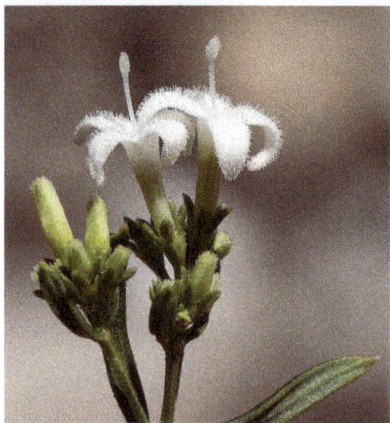

Strophostyles spp.

Wild Bean, Fuzzy Bean

Description Climbing or trailing annual or perennial herbs; leaves alternate, trifoliate, with long petioles, the leaflets with short stalks, linear to broadly ovate, entire on the margin; flowers pea-shaped, pinkish purple, on long peduncles from axils of leaves, developing in summer; pod elongate, several-seeded.

Strophyostyles helvola

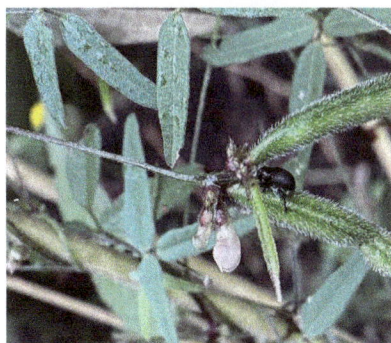

Strophyostyles leiosperma

Distribution Widely distributed.
Habitat Glades, prairies, upland woods or low ground, waste areas, and stream bottoms.

Strophostyles helvola and *Strophostyles leiosperma* are common annuals, the first species identified by glabrous ovate green leaves while the latter species is distinguished by silky-gray foliage and smaller more linear leaflets about 1–2 in. (2½–5 cm.) long.

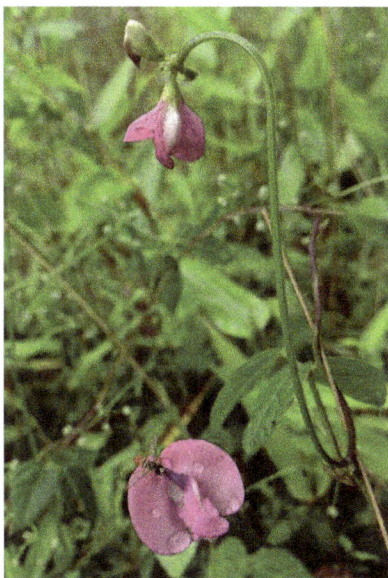

Strophyostyles umbellata

Strophostyles umbellata is a perennial species with narrow-ovate to oblong, somewhat leathery leaflets.

Importance The seeds of the annual wild beans (*Strophostyles leiosperma, S. helvola*) provide food for mourning doves throughout the summer. They are important foods of quail. The plants and seeds make up a small part of the diet of deer. Livestock graze these plants during the summer. The plants are moderately nutritious but usually not present in large enough amounts to contribute much to the total diet.

Pencil-flower

Description Upright perennial herb with usually several wiry stems from a thickened base, to about 20 in. (50 cm.) tall; leaves alternate, trifoliate, with short petiole and conspicuous stipules, the leaflets elliptic, veiny, about 1 in. (2½ cm.) long, with a sharp minute tip; flowers yellow, pea-shaped, developing in late spring and summer; pod short, with hooked tip.

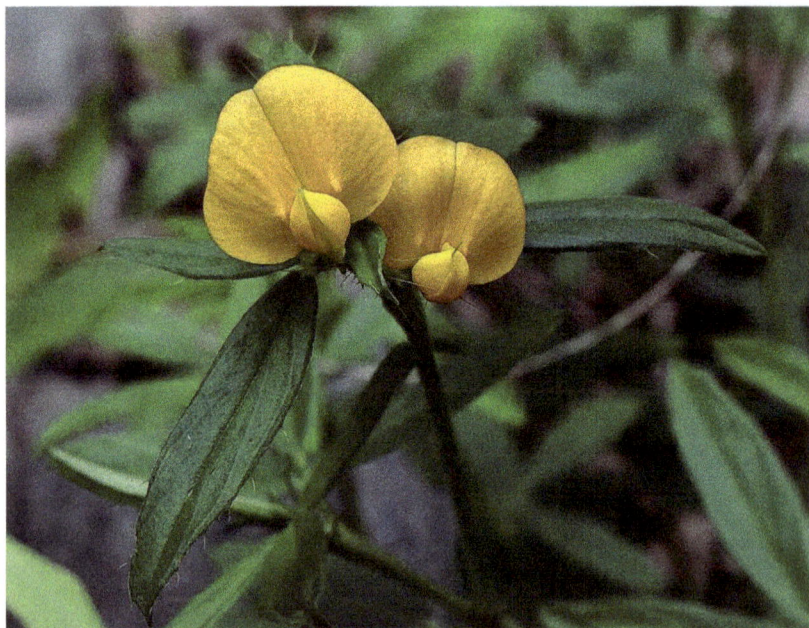

PENCIL-FLOWER, *Stylosanthes biflora*

Distribution Widely distributed.
Habitat Glades and open woods; on dry soils.
Importance Pencil flower is moderately nutritious but only poor to fair forage for livestock. The species is usually not present in large enough quantities to contribute much to the total livestock or wildlife diet. Ruffed grouse make some use of this plant during July. Quail and turkeys occasionally eat the seeds.

Symphyotrichum spp.

Aster

Description A large genus of perennial herbs; stems mostly upright, leafy, to 3 ft. (90 cm.) tall or more; leaves simple, alternate, linear, ovate, or lanceolate, with cordate or tapering base, sessile or petioled; flower heads with pink-purplish, blue, or white rays surrounding a yellow disk, in a leafy simple or branching inflorescence, developing in summer and fall.

Symphyotrichum anomalum

Symphyotrichum cordifolium

Distribution Widely distributed.
Habitat Glades, prairies, open woods, fields, waste areas; also shaded places and low ground for certain species.
Note Many asters are now placed in genus *Symphyotrichum* (and one species described below in *Ionactis*), in part to differentiate North American asters from those of Eurasia.

Two common asters of dry woods habitats with petioled, cordate, or arrow-shaped leaves from base of plant and also on lower stem are *Symphyotrichum cordifolium* (synonym *Aster sagittifolius*) and *Symphyotrichum anomalum* (synonym *Aster anomalus*). The first species

has winged petioles and serrate leaf margins; the second is not winged and has entire margins.

For *Symphyotrichum patens* (synonym *Aster patens*) found on similar sites, the leaves are somewhat cordate but sessile, the base of blade more or less clasping the stem. *Symphyotrichum oblongifolium* (synonym *Aster oblongifolius*) is also sessile, but with small oblong leaves, tapering and usually not clasping the stem.

Symphyotrichum pilosum (synonym *Aster pilosus*) is distinguished by numerous small, linear, fine-pointed leaves and white ray flowers and is common in fields, meadows, and prairies. *Symphyotrichum lateriflorum* (synonym *Aster lateriflorus*), also with white rays but generally coarser leaves, occurs in moist sites and stream bottoms.

Aster

Symphyotrichum patens

Symphyotrichum oblongifolium

Symphyotrichum pilosum

Other common species include **Symphyotrichum turbinellum** (synonym *Aster turbinellus*) with bluish ray flowers and generally glabrous tapering leaves, distinguished from **Symphyotrichum laeve** (synonym *Aster laevis*), by its somewhat clasping leaf bases, both species of dry woods; and **Ionactis linariifolia** (synonym *Aster linariifolius*), with numerous elongate leaves characteristically on acid soils in pine-oak woods.

Other species are present; the most commonly encountered species are described here.

Symphyotrichum spp.

Aster

Symphyotrichum lateriflorum

Symphyotrichum turbinellum

Ionactis linariifolia

Symphyotrichum laeve

Importance The green basal rosettes of asters are eaten by deer through autumn, winter, and early spring; and the stems and flower heads are eaten during summer and early autumn. Asters provide only a very small amount of the total forage eaten by deer but are important because of their widespread occurrence and year-round availability. Turkeys occasionally eat aster flowers and fruits.

The asters provide poor to fair forage for cattle and horses but are generally better for sheep. They are classed as invaders on abused livestock range.

Tephrosia virginiana   FABACEAE, PEA FAMILY

Goat's Rue, Catgut

Description Upright perennial herb with mostly simple or unbranched stems to about 20 in. (50 cm.) tall; leaves alternate, pinnately compound, the leaflets numerous, oblong or elliptic, grayish pubescent, entire on the margins; flowers conspicuous, pea-shaped, yellow with purplish pink, in terminal clusters, developing in summer; pod hairy, several-seeded, maturing in late summer.

GOAT'S RUE, *Tephrosia virginiana*

Distribution Widely distributed.
Habitat Glades, prairies, and open woods; on dry mostly acid soils.
Importance In the Ozarks, goat's rue is considered poor to fair forage for livestock, but in the prairies to the west of the Ozarks it is considered very good forage as it has a high nutritive content, especially in early spring.

Goat's rue has a tough fibrous root system and is a good soil binder for erosion control. The roots are also a source of rotenone, a fish poison.

Clover

Description Erect or decumbent annuals and perennials; leaves alternate, trifoliate, with foliaceous stipules at base, the leaflets minutely serrate on the margin; flowers white, roseate, or yellow, in rounded or oblong heads, appearing throughout the season; pod small, several-seeded.

WHITE CLOVER, *Trifolium repens*

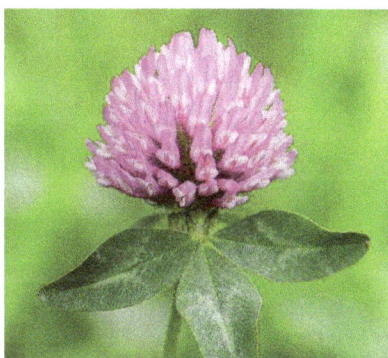

RED CLOVER, *Trifolium pratense*

Distribution Widespread and sometimes common and abundant.
Habitat Most species introduced, naturalizing in fields, pastures, and waste ground.

White clover, *Trifolium repens*, is a low, white-flowered creeping perennial with glabrous runners and generally long-petioled leaves.

Other introduced species include the common **red clover**, *Trifolium pratense*, with roseate or purplish heads and somewhat pubescent stems, and **large hop clover**, *Trifolium campestre* with yellow flowers. **Buffalo clover**, *Trifolium reflexum*, with roseate and white flowers is a native species.
Importance Clovers are highly nutritious and among the most important forage plants in the United States. They are palatable to all kinds of livestock and are readily

eaten, whether green or dry, at all times of the year. Hop clover makes good late-winter and early spring growth, affording nutritious forage when most needed. Red clover and white clover especially are frequently seeded for hay and pasture crops either by themselves or in mixtures with desirable pasture grasses. The clovers all require a high level of available soil nutrients, especially phosphorus, potassium, and calcium, for high yields.

Clover leaves provide important food for prairie chickens during spring and fall, and for ruffed grouse during late winter and spring. Wild turkeys and deer eat small amounts of clover. Quail, doves, and pheasants eat small amounts of the seeds. Rabbits and many smaller herbivorous wildlife species eat clover leaves.

Mullein

Description Erect unbranched biennial from a basal rosette, with soft downy or woolly foliage, and winged stems, to 6–7 ft. (2 m.) tall; leaves simple, alternate, oblong or obovate, entire on the margin; flowers yellow in a spikelike terminal inflorescence, developing in summer.

MULLEIN, *Verbascum thapsus*

Distribution Widespread and sometimes common and abundant.

Habitat Pastures, waste ground, and disturbed areas.

Importance Mullein is a common invader of overused pastures and is a good indicator plant of past overuse. It is poor forage for livestock and is usually eaten only in the late spring when green and then only if more palatable forage is scarce or absent.

Crown-beard, Wingstem

Description Perennial herbs with winged stems, to 3–4 ft. (90–120 cm.) tall; leaves simple, alternate, ovate-lanceolate, with pointed tip and tapering to base, minutely toothed on the margins; flower heads with white or yellow rays, developing in late spring and summer.

FROSTFLOWER, *Verbesina virginiana*

GRAVELWEED
Verbesina helianthoides

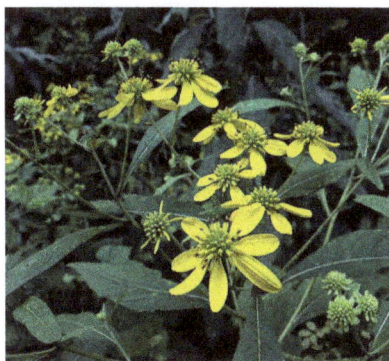

WINGSTEM, *Verbesina alternifolia*

Distribution Widely distributed. **Habitat** Glades, open woods, thickets, and waste ground, also bottomland and alluvium.

Frostflower, *Verbesina virginica*, has short white ray flowers.

Two other species are **gravelweed**, *Verbesina helianthoides*, and **wingstem**, *Verbesina alternifolia*, identified by yellow rays. The first species has 8 or more ray flowers and is found in dry woods. Wingstem has fewer rays and is usually found in moist bottomland sites. **Importance** Crown-beard is eaten occasionally by deer in summer. It is poor forage for cattle but fair for sheep.

Ironweed

Description Coarse perennial herbs with simple leafy upright stems, to 3–4 ft. (90–120 cm.) tall; leaves simple, alternate, narrow-ovate or linear to lanceolate, mostly sessile or with short petioles, finely toothed or nearly entire on the margin; flowers small, purple, in dense heads, lacking ray flowers, developing in late spring and summer.

Vernonia arkansana

Vernonia baldwinii

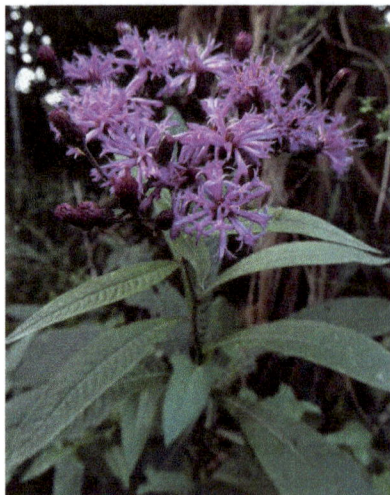

Vernonia gigantea

Distribution Widespread and sometimes common and abundant.
Habitat Glades, prairies, open woods, pastures, low ground, and along small rocky or gravelly streams.

A common species is *Vernonia arkansana* of glades, open woods, and streambanks, with elongate linear leaves less than 1 in. (2½ cm.) wide and finely serrate on the margin.

Two other species are *Vernonia baldwinii* of prairies, open woods, and commonly in pastures, identified by ovate or broadly lanceolate leaves distinctly whitish beneath, and *Vernonia gigantea* of low ground with lanceolate leaves mostly glabrous and greenish beneath.

Importance Ironweed is an undesirable forage plant for livestock and is often abundant where the native range has been overgrazed. A vigorous stand of grasses will control this plant. Mowing or chemical sprays may be necessary.

Vicia caroliniana

Carolina Vetch

Description Spreading or climbing native perennial herb with tendrils, more or less glabrous; leaves alternate, compound, with stipules at base of main leaf stalk, and with more than 10 narrow oblong leaflets, entire on the margins; flowers white and blue in slender racemes on an elongate peduncle, developing in spring; pod about 1 in. (2½ cm.) long.

CAROLINA VETCH, *Vicia caroliniana*

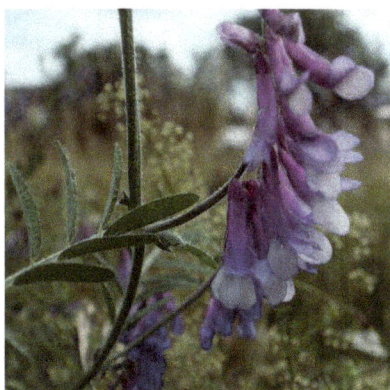

HAIRY VETCH, *Vicia villosa*

Distribution Widely scattered throughout the region, seldom abundant.

Habitat Open woods, dry rocky slopes; on cherty or sandy soils.

The introduced **hairy vetch**, *Vicia villosa*, occurs sporadically, and is easily identified by hairy stems and flower stalks.

Importance Wood vetch is eaten to some extent by livestock, particularly during spring.

Violet

Description Low perennial herbs; leaves simple, from the base of the plant or on the stem, ovate, lanceolate, sagittate, or heart-shaped, or deeply lobed or dissected; flowers spurred, blue, purplish, white, cream-colored, or yellow, or combinations of blue, purple, and white, appearing in spring.

Viola sororia

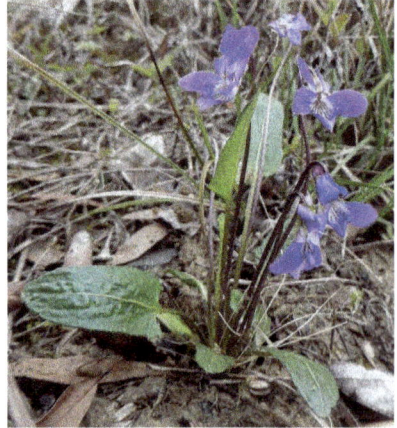

Viola sagittata

Distribution Widespread and sometimes common and abundant.
Habitat Dry to moist woodlands, glades, waste areas, prairies, wet meadows, and bottomlands.

Common violets with blue flowers; and rounded, cordate, or arrow-shaped leaves all from the base of plant include *Viola sagittata* and *Viola sororia*.

Species of damp woodlands and rich soils with ovate or cordate leaves on the stems include *Viola pubescens*, with yellow flowers, and *Viola striata*, with whitish flowers.

Bird's-foot violet, *Viola pedata*, with deeply dissected basal leaves and pansy-like flowers, is common on dry rocky soils.

Wild pansy or **Johnny-jump-up**, *Viola bicolor*, is a delicate annual with small stem leaves and dissected stipules, and small, bluish or whitish, pansylike flowers.
Importance The green overwintering basal leaves of violets provide occasional food for deer. Violets provide food for game birds including ruffed grouse, mourning doves, prairie chickens, and pheasants.

Viola spp.
Violet

Viola pubescens

Viola striata

BIRD'S-FOOT VIOLET, *Viola pedata*

WILD PANSY, *Viola bicolor*

Cocklebur

Description Coarse branching annual with thick stems to 3 ft. (90 cm.) tall; leaves simple, alternate, scabrous, ovate or cordate with toothed margins; flowers small, inconspicuous, clustered, developing in summer; bur ovoid, up to 1 in. (2½ cm.) long, with hooked prickles and 2 seeds.

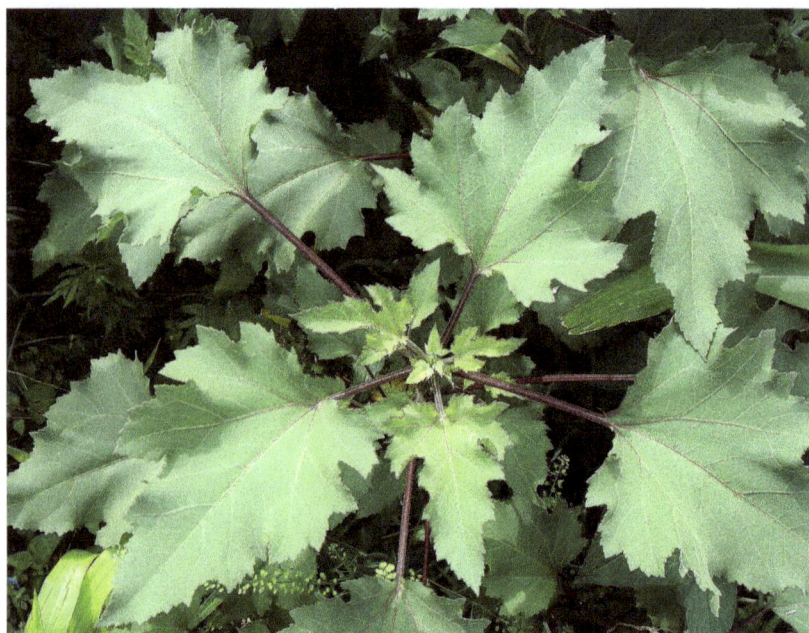

COCKLEBUR, *Xanthium strumarium*

Synonym *Xanthium pensylvanicum*
Distribution Widespread and sometimes common and abundant; a noxious weed in Arkansas and many other states.
Habitat Fields, waste ground, stream bottoms, and overflow areas.
Importance Cocklebur is worthless to poor as a forage plant for livestock, being eaten only in very early spring when other forage is scarce.

Hogs are frequently poisoned by this plant, and it may be fatal to all kinds of livestock. Germinating seeds and cotyledons of young plants are most toxic. Poisoned animals usually show weakness, unsteady gait, labored respiration, weak pulse, low temperature, and sometimes nausea and vomiting. The symptoms may appear within 24 hours after eating the plant.

Mowing the plants before seed formation and spraying with an herbicide are generally effective control measures.

Grasses and Grasslike Plants

G rasses and grasslike species are herbaceous plants with slender stems, generally elongate leaves with parallel veins, and small flowers. True grasses belong to the family Poaceae. Grasslike plants are the sedges and rushes belonging to the families Cyperaceae and Juncaceae, respectively.

Although the importance of grasses for livestock is well known, their value for wildlife is sometimes overlooked. For instance, many annual grasses produce large crops of grain or "seeds" (technically *caryopses*), which provide food for numerous wildlife species. This section describes and indicates the distribution of each grass and grasslike plant, and also discusses its importance as food for wildlife and livestock.

Because identification of plants found in dropping, stomach, or crop samples is difficult, the literature frequently specifies foods only as "unidentified grass." For some species of wildlife, these unidentified grasses are major components of their diet. Since this use does not appear under any plant species discussions, it is mentioned briefly below.

Grass and sedge leaves are eaten heavily by wild turkeys in winter and spring and comprise about 25 percent by volume of their year-round diet; only acorns make up a greater volume. Deer eat more grass and sedge leaves than any other food in spring, a considerable amount in winter, and some leaves throughout the year. Prairie chickens also eat grass and sedge leaves, about five percent of the total diet, throughout the year. Pheasants, coyotes, foxes, and ruffed grouse eat smaller amounts, and most wildlife species, intentionally or accidentally, include some green grass or sedge leaves in their diets. Greatest amounts are eaten during spring when grass plants are tender and succulent.

The vegetative morphology of grasses and grasslike plants includes certain characteristics that distinguish true grasses from sedges and rushes (TABLE 1). Rootstock, stem, and leaf characters provide a means of field recognition, but whenever possible the flowering stalk and spikelets should also be examined as they are sometimes necessary for accurate identification.

The **ligule** is one of the most useful vegetative characteristics for identification. It is either a collarlike membrane of various textures, shapes, and lengths, or a fringe of ciliate hairs at the inner junction of the leaf blade and sheath. **Auricles** are two minute appendages or lobes at the junction of the sheath and leaf blade. These structures occur primarily in the barley tribe. **Vernation** refers to the rolled or folded arrangement of the young grass leaf still inside the sheath. To determine vernation, a cross section of the stem is made through the sheath of the uppermost fully expanded leaf.

Some perennial grasses may spread by underground rhizomes or by surface runners called stolons. These are modified stems that develop roots at the nodes, thereby forming new plants and developing a turf or sod. Kentucky bluegrass, quackgrass, and Johnson grass are examples of species propagating by rhizomes. Bermuda grass spreads by stolons and rhizomes. In some grasses that form tufts or bunches, such as big bluestem, short rhizomes may be present.

TABLE 1. General comparisons between grasses (Poaceae), sedges (Cyperaceae), and rushes (Juncaceae).

CHARACTER	POACEAE	CYPERACEAE	JUNCACEAE
Stem or culm	Mostly hollow, and rounded to somewhat flattened	Mostly solid or pithy, and 3-angled	Somewhat pithy, and rounded
Node	Conspicuous	Indistinct	Indistinct
Leaf arrangement	Two-ranked	Three-ranked	Two-ranked
Leaf sheath	Split or overlapping, but mostly closed closed in some species	Mostly closed	Open or closed (open in rush)
Ligule	Present in most species	Absent or indistinct	Absent or indistinct
Auricles	Mostly absent, but present in some species, such as wildrye	Absent	Present or absent
Leaf blade	Flat or V-shaped	Flat or V-shaped	Flat or round

Bentgrass, Redtop

Description Principal species perennial, developing tufts, or sod-forming and spreading by rhizomes, with stems to 3 ft. (90 cm.) tall; leaf blades about 3/16 in. (5 mm.) wide or less; ligules whitish, conspicuous; vernation rolled; panicles erect with spreading branches and 1-seeded spikelets.

REDTOP, *Agrostis stolonifera*

Distribution Widespread and sometimes common and abundant.
Habitat Fields, pastures, and woodlands; on dry to moist soils; the common redtop introduced from Europe as a cool-season forage plant.

Redtop, *Agrostis stolonifera,* spreads by rhizomes and forms a loose turf or sod; leaf blades long, flat, 3/16 in. (5 mm.) wide; sheaths shorter than the internode; ligule about ⅛ in. (3–4 mm.) long; panicles stiffily erect, with whorls of spreading branches, reddish purple, developing in late spring.

Ticklegrass, *Agrostis hyemalis*, is a native perennial of prairies and open woods, forming dense tufts, flowering in spring.

Autumn bentgrass, *Agrostis perennans*, flowering later in summer, is a native perennial species of damp or shaded sites, identified by loose tufts and somewhat decumbent or weak-reclining stems.

Bentgrass, Redtop

TICKLEGRASS, *Agrostis hyemalis*

Importance Redtop, the most important species of this genus, is a cool-season plant. On moist sites, however, the foliage remains green throughout the summer and is readily grazed by livestock until August or September when seed is mature. Although redtop is not considered as palatable as some introduced grasses used for reseeding, it is commonly mixed with desirable legumes for both pasturage and hay. It is also used in lawn mixtures and is a valuable soil stabilizer in eroded areas because of its vigorous growth and good turf-forming ability.

Hairgrass and autumn bentgrass, two cool-season native perennials, are less palatable than redtop but

TICKLEGRASS, *Agrostis hyemalis*

are grazed fairly readily by livestock until flower stalks develop.

All the species mentioned are considered increasers on native rangelands.

Bluestem, Broomsedge, Turkeyfoot

Description A genus with 200 or more species on a world basis; bunch-forming, medium-sized to tall perennials; foliage glabrous to heavily pubescent; leaf blades ⅛–⅜ in. (3–10 mm.) wide; sheaths somewhat compressed in the tuft; ligules collar-shaped with fringed margin; vernation rolled or folded; flower stalks terminal or from axillary sheaths, consisting of fragmenting racemes of paired spikelets, one sessile, awned, the other stalked, awnless or with short awn, developing in middle to late summer.

BIG BLUESTEM
Andropogon gerardii

LITTLE BLUESTEM
Schizachyrium scoparium

Distribution Widely distributed, sometimes locally abundant and forming dense stands.

Habitat Glades, prairies, open woods, old fields, and waste ground.

Little bluestem, *Schizachyrium scoparium* (synonym *Andropogon*

scoparius), is a common prairie and glade species with erect culms to 3 ft. (90 cm.), leaf blades fine, about 3/16 in. (4–5 mm.) wide, glabrous to densely villous and bluish green in some plants, particularly in the southwestern part of the Ozark range, and with delicate exserted racemes.

Big bluestem or **turkeyfoot**, *Andropogon gerardii*, is a taller coarser species growing to 6–7 ft. (2 m., sometimes more), with short rhizomes; flowering stalk erect, with several digitate or fingerlike racemes.

Bluestem, Broomsedge, Turkeyfoot

BROOMSEDGE, *Andropogon virginicus*

A common species of old fields and abandoned ground is **broomsedge**, *Andropogon virginicus*, forming tufts, with conspicuously keeled lower sheaths; racemes feathery in congested fascicles from upper sheaths; plants tawny to reddish brown in winter condition.

A less extensive species, primarily of western and southern parts of the Ozarks, is **silver beardgrass**, *Bothriochloa laguroides* (synonym *Andropogon saccharoides*), with somewhat bluish-green foliage and fluffy silvery terminal panicles.

Importance Little bluestem is one of the most abundant and generally distributed range plants in the Ozarks. A warm-season midgrass, it is moderately palatable and eaten by all kinds of livestock, especially in the early stages of growth. It begins growth in April and provides forage of adequate nutritive value for most kinds of livestock until mid-July when it begins to produce seedstalks. After maturity it is not as readily eaten by livestock but makes fair fall and winter grazing for cattle and horses when supplemented with protein and minerals. This grass can produce up to 2 tons of cured forage per acre. Little bluestem is a decreaser on native ranges and, when continually grazed closely during the growing season, is replaced by less productive species.

Big bluestem is one of the most important native grasses in the Ozarks. This is a tall warm-season grass and can produce a large

Andropogon spp.

Bluestem, Broomsedge, Turkeyfoot

BROOMSEDGE
Andropogon virginicus

SILVER BEARDGRASS
Bothriochloa laguroides

volume of forage on a small area of ground. The forage, especially when young and green, is highly relished by cattle and horses. Growth begins in April and provides forage of high nutritive quality until mid- to late-July. At maturity, the seedstalks become hard and coarse, but the rest of the plant furnishes fair fall and winter grazing for cattle and horses when the animals are fed a protein supplement. Big bluestem occasionally has been seeded in pure stands and can produce 1–2½ tons per acre of high-quality hay. Big bluestem, like little bluestem, decreases on native ranges when continually grazed closely during the growing season.

Silver beardgrass is a native grass of the western Ozarks. It is aggressive, rapidly occupies disturbed areas, and is considered an invader in the true prairie. It is less palatable than either big or little bluestem.

Broomsedge is a poor forage plant of moderate to low nutritive value. It is a common invader on overused native ranges, old fields, and low-fertility tame pastures. This warm-season native perennial grass is occasionally eaten by cattle in early spring if more palatable grasses are not present. However, if the previous year's seedstalks are not removed by burning or mowing, even the new spring growth is seldom taken. The old dried growth is rarely grazed.

Dense stands of Andropogon provide winter cover especially valuable for prairie chickens and quail. Seeds serve as minor foods for quail, deer, and prairie chickens. Turkeys occasionally eat the young leaves of little bluestem.

Poverty Grass, Three-awn Grass, Triple-awn

Description Annuals and perennials forming thin tufts, lacking rhizomes, with slender branching stems usually not exceeding 20 in. (50 cm.); leaf blades narrow, about 1/16 in. (1½ mm.) wide; ligules minute, less than 1/32 in. (½ mm.) long; vernation rolled; panicles narrow to somewhat loose and flexuous, the spikelets characteristically three-awned, with sharp-pointed base, developing in summer.

PRAIRIE THREE-AWN
Aristida oligantha

ARROW-FEATHER THREE-AWN
Aristida purpurascens

Distribution Scattered throughout the region, sometimes locally common.

Habitat Open woods, fields, waste areas; on thin dry soils.

Prairie three-awn, *Aristida oligantha,* one of the most common annuals, is distinguished by loose flexuous panicles and awns about 1½ in. (3–4 cm.) long, all approximately the same length.

Poverty grass, *Aristida dichotoma,* another annual, is similar but readily distinguished by its shorter strongly unequal awns, the two lateral awns only about ⅛ in. (2–3 mm.) long.

A third common species of dry Ozark habitats, **arrow-feather three-awn,** *Aristida purpurascens,* is a tufted perennial with long narrow panicles one-half the length of the entire plant and spikelets with subequal awns, the two laterals usually ⅜ in. (1 cm.) or longer.

Importance Quail eat minor amounts of *Aristida* seeds, and use mature grass stands for roosting cover. Three-awns are considered poor forage for livestock in the Ozarks, although they are grazed to some extent in the early growth stage when green and succulent. The barbed seeds of mature plants often trouble animals by working into their eyes, nostrils, and ears, and sometimes cause sore mouths.

Arundinaria gigantea

Giant Cane

Description Coarse perennial, sometimes forming dense colonies, spreading by rhizomes, the only genus of woody grasses in the Ozark region; stems or culms bamboolike, mostly leafy or occasionally leafless, growing to 20 ft. (6½ m.) tall or occasionally more but generally shorter in Ozark habitats; leaf blades lanceolate with constricted base or short petiole; sheaths hispid, overlapping each other; ligule minute; vernation rolled; seldom flowering or producing seed.

GIANT CANE, *Arundinaria gigantea*

Distribution Scattered throughout the central and southern Ozarks.
Habitat Low woods, streambanks, spring banks, and alluvial bottoms.
Importance The plant commonly grows in small colonies or thickets and can be very valuable forage, especially since the foliage usually remains green throughout the year. Giant cane is nutritious and palatable to livestock, which readily eat the young plants and the seeds and leaves of older plants. It may also be eaten to some extent by deer in winter when green forage is scarce.

Giant cane is a decreaser and is easily destroyed by continuous heavy grazing and rooting of swine. It undoubtedly was more extensive at one time but has been reduced by cultivation, burning, and overgrazing; it comes back well if moderately grazed and protected from fire.

The culms are used for fishing rods, pipe stems, baskets, mats, and many other purposes.

Sideoats Grama

Description Perennial, developing leafy tufts, with culms generally to 20 inches (50 cm.) tall; leaf blades narrow, ⅛ in. (3–4 mm.) wide, in-rolling when dried, and tapering to stringlike tips; ligule somewhat ciliate; vernation rolled; flowering stalks erect, more or less 1-sided with numerous pendulous spikes about ½ in. (10–12 mm.) long, developing in early summer to midsummer.

SIDEOATS GRAMA, *Bouteloua curtipendula*

Distribution Scattered throughout the region, seldom abundant.
Habitat Dry prairies, limestone glades, woods openings, and bluffs; mostly on neutral or alkaline soils.
Importance Sideoats grama is a valuable forage plant wherever it is abundant. This native warm-season bunch grass is highly palatable to all kinds of livestock. Although grazed mostly during the growing season, it also provides fair winter forage. It usually starts growth in April and is a vigorous grower.

Sideoats grama is considered an increaser and will usually replace the taller grasses, such as big bluestem, on overused ranges; it will decrease if closely grazed during the growing season.

Bromus spp.

Brome, Cheat, Chess

Description Annual or perennial grasses, sometimes loosely tufted, with upright culms 20–40 in. (50–100 cm.) tall or more; leaf blades ⅛–⅝ in. (3–15 mm.) wide; sheath closed; ligule conspicuous, notched or toothed on margin; vernation rolled; panicles terminal, with mostly awned several-flowered spikelets, developing in spring. Four species included here are introduced annuals of open ground, the fifth a native perennial of woods habitats.

CHEATGRASS, *Bromus tectorum*

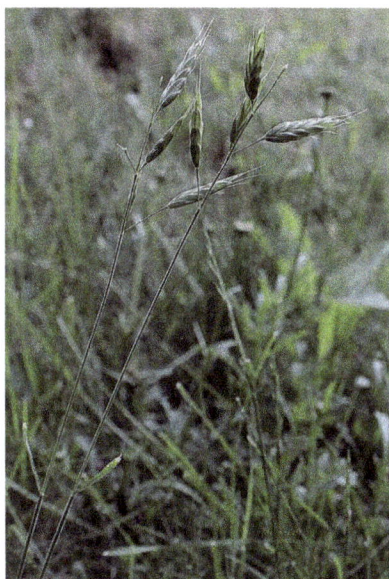

HAIRY CHESS, *Bromus racemosus*

Distribution Widespread and sometimes common and abundant.
Habitat Fields, pastures, and waste areas; on dry ground or shaded habitats; includes several introduced annuals from Europe and Asia.

A distinctive introduced annual, and one of the most common, is cheatgrass, *Bromus tectorum*, with pubescent foliage and silvery green panicles; spikelets narrow, long-awned from pendulous branches, one of the earliest to flower in spring.

Two other annuals, also pubescent, are **Japanese chess, *Bromus japonicus*** (sometimes considered synonymous with *Bromus tectorum*), and **hairy chess, *Bromus racemosus***, both with shorter awns generally less than 25 in. (1 cm.), the first species with diffuse spreading panicles, the latter usually more compact.

An annual with smooth foliage and short-awned, 15 in. (5 mm.), or occasionally awnless spikelets is **common chess, *Bromus secalinus*.**

Brome, Cheat, Chess

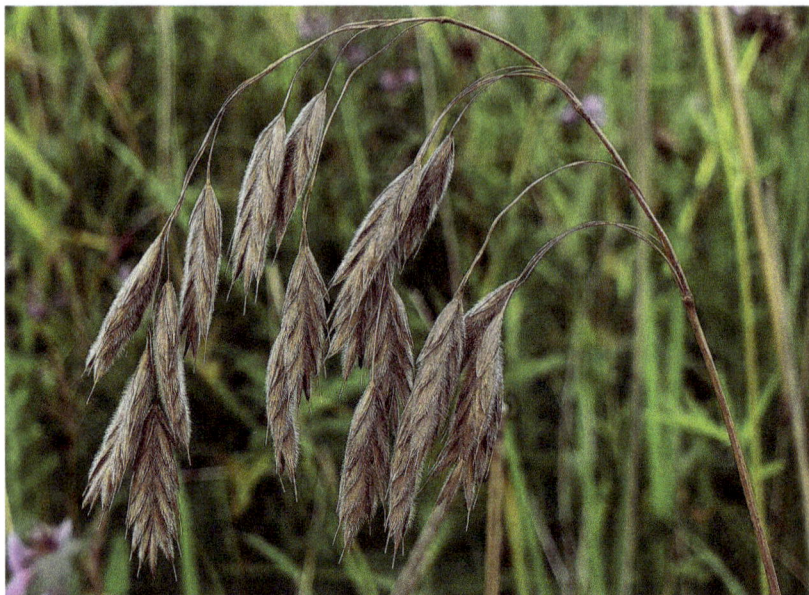

CANADA BROME, *Bromus kalmii*

Canada brome, *Bromus kalmii*, is a native perennial of woodlands, solitary or forming sparse clumps, to 40 in. (100 cm.) tall or more, with pubescent foliage and drooping panicles, the spikelets also generally pubescent and distinctly awned.

Importance Reports for the Ozarks show light use of *Bromus* spp. by deer during the winter, turkeys during spring, summer, and autumn, quail in spring, prairie chickens year round, ruffed grouse during summer, and raccoons. Pheasants utilize brome in northern Missouri.

In general, the chess grasses are not palatable to livestock, but these cool-season grasses are frequently the only green succulent plants available in late February and March and are then grazed quite heavily. The grasses may germinate in autumn, remain green in winter, and produce seed in May or June. They give little competition to the warm-season grasses because they begin growth when the warm-season grasses go dormant in the fall and make maximum growth in the early spring. Downy chess in particular is a noxious weed, because its sharp awns may injure the eyes and nostrils of animals. This grass frequently grows in dense stands and is a fire hazard when dry.

Canada brome is considered an increaser in overgrazed wooded ranges. The annual chess grasses are invaders in abused old fields, pastures, and native ranges; they may be easily controlled by range management practices that yield a vigorous stand of more palatable perennial grasses.

Carex spp.

Sedge

Description A complex genus of the sedge family with numerous species on a world basis, the carices are perennial grasslike plants distinguished from grasses (Poaceae) by leafy triangular stems, closed sheaths, and minute reduced flowers in saclike bracts (perigynia), forming loose to somewhat compact clusters in a terminal inflorescence. Five widely prevalent and characteristic species are described for the Ozarks.

Carex albicans

Distribution Distributed throughout the region, seldom abundant.

Habitat On various sites from dry open woods, clearing, and fields, to wet woods, calcareous swales, and alluvial soils.

A most common sedge of upland woods on dry soils is *Carex albicans* with tufts of narrow leaves, reddish toward base, and slight stems about 10 in. (25 cm.) tall or less. It is among the earliest carices to resume growth and flowering in spring.

Loosely tufted and coarser species of similar sites are *Carex hirsutella*, and *Carex muehlenbergii*. The first species has lax pubescent foliage, the second somewhat stiff harsh leaves.

A conspicuous species of woods and clearings, usually wintergreen with thick bluish-green foliage, is *Carex glaucodea*.

Carex lurida, of low wet prairies, seeps, and stream bottoms, is a tall species growing to 3 ft. (90 cm.) or more, with long, lax, yellow-green foliage, and orange rootstocks.

Many more species of *Carex* are present in the Ozark region, in both dry and wet habitats, and positive identification typically requires mature floral parts, especially the perigynia.

Sedge

Carex hirsutella

Carex muehlenbergii

Carex glaucodea

Carex lurida

Importance Sedges are important wildlife foods. Leaves and seeds (achenes) are eaten throughout the year by wild turkeys, seeds and floral parts are important in the spring diet of prairie chickens, seeds are taken throughout summer by ruffed grouse, and some use by quail and pheasants has been noted. Deer graze sedge leaves during winter and spring.

The foliage ranks from fair to very good as forage for cattle, sheep, and goats.

Some sedges remain green and succulent during winter provide a valuable source of forage when most plants are dormant.

Spikegrass, Wood-oats

Description Leafy perennial, forming loose sparse clumps from short rhizomes; culms erect, to about 40 in. (1 m.) tall; leaf blades glabrous, to 1 in. (2½ cm.) wide, with conspicuous venation; ligule minute, ciliate; vernation rolled; panicles showy, nodding, with broad strongly flattened spikelets as much as ⅝ in. (15 mm.) wide, developing in summer and later turning brown.

SPIKEGRASS, *Chasmanthium latifolium*

Synonym *Uniola latifolia*
Distribution Widely scattered throughout the region, seldom abundant.
Habitat Alluvial bottoms, creekbanks, and wooded areas.
Importance This woodland species is only moderately palatable and generally not readily grazed by livestock unless better quality forage is scarce or absent.

Deer occasionally graze this plant, and quail and turkeys eat small amounts of the seeds.

It is sometimes planted as an ornamental.

Bermuda Grass

Description Low perennial, sod-forming, spreading by scaly runners, stolons, or rhizomes, with leafy culms to about 15 in. (35–40 cm.) tall; leaf blades short, about 3/16 in. (3–4 mm.) wide; several strains with fine to coarse-textured foliage; ligules a conspicuous fringe of hairs; vernation folded; inflorescence digitate, consisting of several short spikes from top of flowering stalk, developing in summer; spikelets 1-flowered, on one side of the rachis.

BERMUDA GRASS, *Cynodon dactylon*

Distribution Sporadic, becoming more common southward.
Habitat Fields, pastures, and waste ground; introduced from the Old World for turf and pasture use.
Importance Bermuda grass is one of the most important pasture grasses in the southern United States. This warm-season grass is found throughout the Ozarks. It is highly palatable to all kinds of livestock and rates as an excellent forage plant. Several improved strains of Bermuda grass are used in pasture reseeding. It is also used extensively as a lawn grass in many areas.

This grass is aggressive and spreads rapidly when cultivated; but if not renovated it will become sod-bound and lose vigor in a few years.

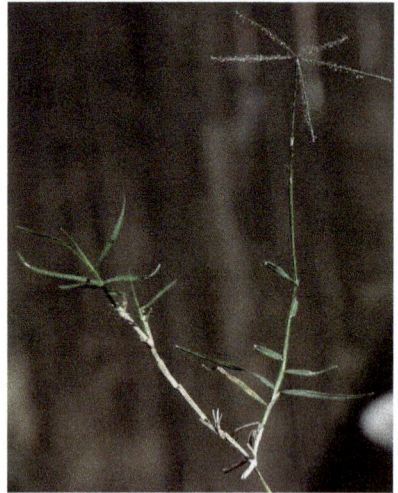

Although the plant will grow in poor soils, it requires fertile conditions to produce a large volume of forage. It responds very well to fertilizers high in nitrogen.

Cyperus esculentus

Chufa, Yellow Nutgrass

Description Smooth perennial with underground runners producing small tubers; stems triangular, lacking the nodes or joints found in grasses, 10–20 in. (25–50 cm.) tall or more, leafy at base; foliage yellowish green; umbellate clusters of flat yellowish spikelets at summit of flowering stalk, subtended by straplike leaves, developing in early summer.

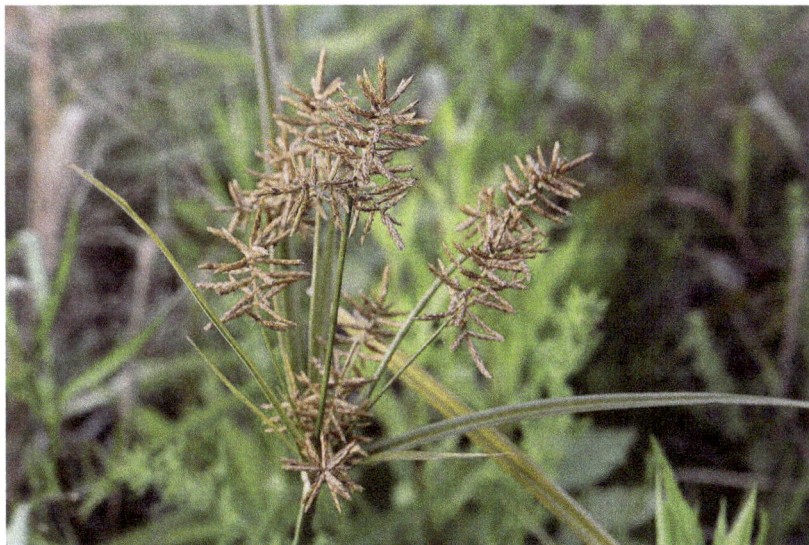

CHUFA, *Cyperus esculentus*

Distribution Distributed throughout the region, seldom abundant.
Habitat Wet fields, sand and gravel bars, swales, and borders of sloughs, ponds, and streams.

Another common native species is **Cyperus strigosus**, similar in general appearance but without tuber-bearing stolons.
Importance The "seeds" (achenes) and underground tubers of chufa are favored by ducks and Canada geese. Quail make light use of these plants, as do deer, turkeys, prairie chickens, pheasants, and ruffed grouse.

Chufa is considered a poor forage plant for cattle, sheep, and goats. Hogs, however, are fond of the nutritious tubers (below).

Orchard Grass

Description Leafy perennial forming dense tufts, with erect culms to about 3 ft. (90 cm.); leaf blades glabrous to somewhat scabrous, elongate, ⅜ in. (10 mm.) wide; ligule quite conspicuous, thin, whitish, about 3/16 in. (5 mm.) long; vernation folded; panicles erect, with thick 1-sided clusters of spikelets from short stiffly ascending or spreading branches, developing in late spring and early summer.

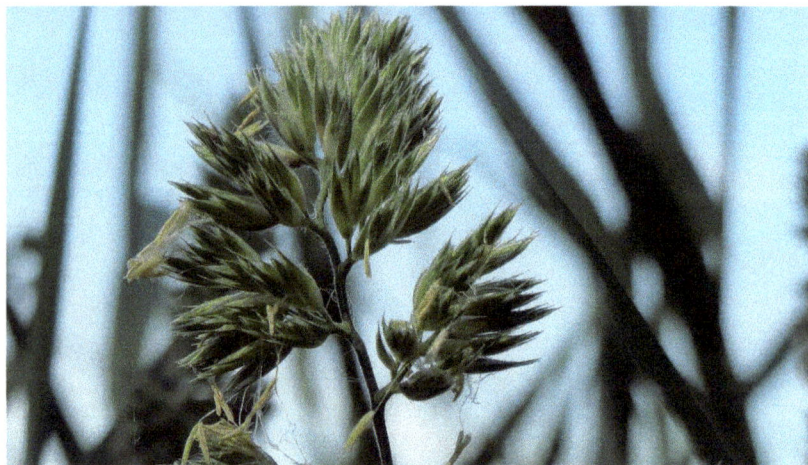

ORCHARD GRASS, *Dactylis glomerata*

Distribution Scattered throughout the region, occasionally abundant.
Habitat Pastures, meadows, fence rows, and similar habitats; introduced from Europe for forage.
Importance Orchard grass is one of the most important and widely used pasture grasses in the Ozarks. This long-lived, cool-season bunch grass produces abundant forage relished by all kinds of livestock. This grass is particularly valuable because it starts growth very early in the spring, characteristically makes fall regrowth, and may remain green well into the winter months.

Deer occasionally crop it in late winter and early spring when green forage is scarce. Orchard grass is grazed by wild turkeys in winter and provides cover and a source of insects for young turkey poults in late summer. Canada geese feed upon the seeds and leaves, and rabbits use the plant for both food and cover.

Orchard grass is highly favored for reseeding pastures, usually in combination with legumes, is adaptable to a wide variety of soils, and has growth characteristics which make it useful both as a hay crop and as pasturage. Although orchard grass grows on poor soils, it does best on moist rich soils and should be fertilized, particularly with nitrogen and phosphorus, for optimal production.

Poverty Oatgrass

Description Low perennial of the oat tribe forming clumps with flowering stalks to about 20 in. (50 cm.) tall; leaves curling, the blades narrow, stringlike, mostly glabrous or somewhat pilose; ligule a fringe of hairs; vernation folded; panicles narrow with few scattered spikelets and conspicuous pointed glumes (pair of bracts at base of each spikelet), developing in late spring.

POVERTY OATGRASS, *Danthonia spicata*

Distribution Scattered throughout the region, sometimes locally abundant.

Habitat Glades, open woods, and clearings; on dry thin soils.

Importance Poverty oatgrass has a high shade tolerance and is common in overgrazed and burned forest ranges. This cool-season bunch grass produces a relatively small amount of foliage per plant and is generally of low nutritive value.

Poverty oatgrass is eaten by cattle and wild turkeys for a short time in early spring and late autumn when other green forage is scarce. It is seldom eaten from late spring or early summer until autumn when growth is resumed. Deer graze the plant in very limited amounts.

Poverty oatgrass increases on abused forest ranges and is considered an indicator of very poor soil.

Crabgrass

Description Tufted annuals, somewhat decumbent, rooting at lower nodes and spreading or more upright, mostly less than 20 in. (50 cm.) tall; leaf blades of various widths: ligule collarlike; vernation rolled; inflorescence digitate, consisting of narrow spikelike racemes from terminal part of flowering stalk, with spikelets 1-flowered, arranged on one side of the rachis, developing in summer.

SMOOTH CRABGRASS
Digitaria ischaemum

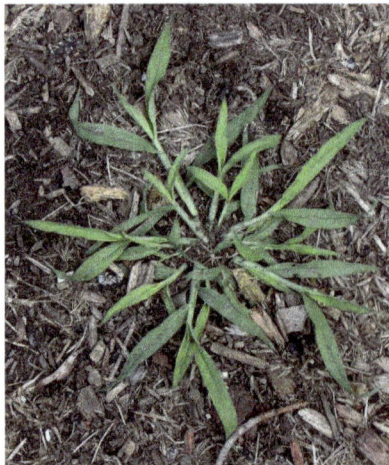

HAIRY CRABGRASS
Digitaria sanguinalis

Distribution Widespread and sometimes common and abundant.
Habitat Fields, cultivated ground, lawns, waste areas, glades, prairies, and open woods. Two of the three species described are naturalized from the Old World.

Two common introduced species are **smooth crabgrass**, *Digitaria ischaemum*, and **hairy crabgrass**, *Digitaria sanguinalis*. Both rooting at the nodes and spreading; rachis flat, about 1/16 in. (1 mm.) wide; the first species has smooth foliage and short ligule about 1/16 in. (1½ mm.) long, and the latter has soft pilose hairs and longer ligule, ⅛ in. (3 mm.).

The native **fingergrass**, *Digitaria filiformis*, of glades and open woods is more erect with narrow leaves and slender racemes, the rachis somewhat triangular instead of flat and only about ½ mm. wide.

Digitaria spp.
Crabgrass

FINGERGRASS, *Digitaria filiformis*

Importance Crabgrass seeds and leaves are especially important for wild turkeys in the Ozarks. They are eaten every month of the year. Lighter use by doves, quail, Canada geese and ducks, prairie chickens, pheasants, ruffed grouse, and raccoons has been reported. The seeds are considered an important food for many songbirds.

Although seldom eaten in the dried stage, crabgrasses are moderately palatable to livestock when green, particularly late in the grow-

ing season when other green forage may become scarce. At that time, they can produce a large amount of forage per plant if moisture conditions are suitable.

The crabgrasses are considered invaders on abused native ranges and pastures and are a nuisance in lawns and cultivated ground. The low growth form, ability to root at the nodes, and heavy seed production make these grasses difficult to eradicate except by chemical means.

Barnyard Grass, Water Grass, Wild Millet

Description Coarse annuals, somewhat decumbent, branching from lower nodes, to as much as 4½ ft. (1½ m.) tall; foliage mostly glabrous or slightly scabrous, the leaf blades to ⅜ in. (1 cm.) wide; ligule absent; vernation rolled; panicles bristly, erect or somewhat nodding, usually heavy with thick spikelike racemes, the spikelets plump, awned or awnless, developing in middle and late summer.

Echinochloa muricata

Echinochloa crus-galli

Distribution Widespread and sometimes common and abundant.

Habitat Fields, waste areas, on damp ground, borders of ponds and lakes, alluvial flats, and river bottoms; barnyard grass, *Echinochloa crus-galli*, introduced from Europe and now widely naturalized.

Two species closely similar in general appearance and requiring careful diagnosis of fruiting characters for separation are *Echinochloa crus-galli*, an introduced species, and the native *Echinochloa muri-*

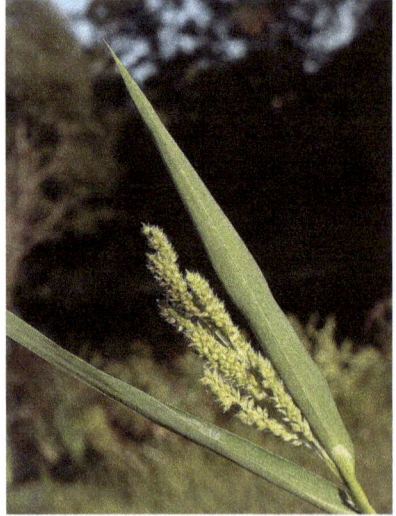

cata. Long-awned varieties occur for both species.

Importance These plants are heavy seed producers and are an important food for Canada geese and ducks. Use by doves, pheasants, ruffed grouse, and raccoons, is recorded for the Ozarks.

These warm-season annuals are nutritious and moderately palatable to all kinds of livestock. They are not dependable forage plants, however, and have never become important for grazing by livestock.

Barnyard grass is considered an invader on many ranges and a weed in cultivated fields.

Elymus spp.

Wild-rye

Description Tufted perennials of the barley tribe, with leafy culms 2–4 ft. (0.6–1.1 m.) tall or sometimes more; leaf blades 1–4 in. (5–20 mm.) wide; auricles generally conspicuous; ligules small, collar-shaped; vernation rolled; spikes bristly, stiff, erect to somewhat nodding, developing in early summer.

Distribution Widely scattered throughout the region, sometimes locally abundant.

Habitat Low ground, streambanks, shaded woods; also upland prairies and swales.

A common species of damp woodlands is **Virginia wild-rye**, *Elymus virginicus*, with stiff straight spikes sometimes partially enclosed in the upper sheaths, with straight awns or awnless. *Elymus villosus* is also a woodland species, recognized by soft pubescence on upper surface of leaf blades, and pubescent, somewhat nodding spikes with straight awns.

Canada wild-rye, *Elymus canadensis*, of prairies and glades is a coarse species to 4½ ft. (1.3 m.) tall or more, with green to distinctly bluish-green foliage and coarse nodding spikes with long spreading or divergent awns.

Another *Elymus*, **quackgrass** (*Elymus repens*), is discussed under *Pascopyrum*.

Importance The wild-ryes are palatable and provide nutritious forage to all kinds of livestock when young and green, but they become coarse and tough when dried and mature. These winter-hardy grasses begin growth in early autumn and remain green into winter.

VIRGINIA WILD-RYE
Elymus virginicus

They should be harvested early because the seedheads may become infected with ergot, a fungus disease. The fungus attacks the seedheads and replaces the seed with hardened black or purplish kernel-like structures. Ergot has cumulative effects and may severely poison cattle, sheep, and horses.

The wild-ryes are decreasers on native ranges and, because of their palatability, are sensitive indicators of overgrazing.

Stinkgrass

Description Low tufted annual, with leafy culms, mostly less than 20 in. (50 cm.) tall; margins of leaf blades, sheaths, and nodes with minute glands (observed with hand lens); ligule minute, consisting of hairs; vernation rolled; panicles short-branching, with numerous, linear, many-flowered, somewhat compact spikelets, developing in early summer.

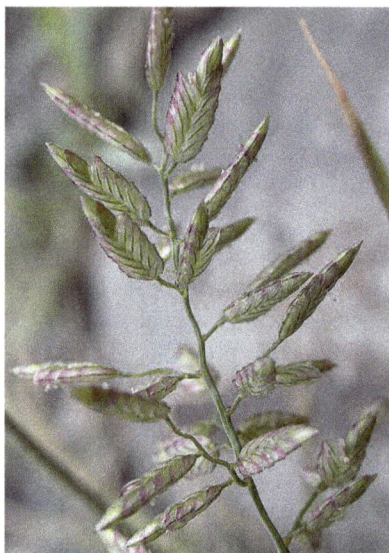

STINKGRASS, *Eragrostis cilianensis*

Distribution Widespread and sometimes common and abundant.
Habitat Fields, cultivated ground, pastures, and waste areas; naturalized from Europe.
Importance Stinkgrass, a warm-season grass with a very disagreeable odor when fresh, has a low palatability rating and is eaten by livestock only when young and green and when good forage is not available. It is a common invader on abused ranges and pastures and can be controlled by proper range management.

This grass is capable of poisoning livestock, especially horses, if eaten in large amounts.

Festuca spp.

Fescue

Description Annual or perennial grasses, short, slight, to tall, robust, and forming large tufts; ligule minute, less than 1/32 in. (½ mm.) long; vernation rolled; auricles sometimes present; panicles terminal with short ascending to somewhat spreading branches and several-flowered spikelets, developing in spring.

Distribution Widespread and sometimes common and abundant.
Habitat Fields, meadows, prairies and woodland, or waste ground; on dry or moist soils; **tall fescue**, *Festuca arundinacea* (now often classified as *Schedonorus arundinaceus*), introduced from Europe.

Six-weeks fescue, *Festuca octoflora* (synonym *Vulpia octoflora*), is a common native annual of dry sterile ground, less than 20 in. (50 cm.) tall, with narrow glabrous leaves about 1/16 in. (1–2 mm.) wide; panicles upright, narrow, with delicate many-flowered spikelets, awned (sometimes awnless in var. *glauca*).

Kentucky 31 is an important strain of tall fescue discovered in this country, distinguished by coarse glossy green foliage and large dense clumps; auricles present with cilia on the margins; panicles narrow, with ascending branches; spikelets awnless or shortawned.

Two native perennials are **nodding fescue**, *Festuca subverticillata*, and *Festuca paradoxa*, solitary or loosely tufted, the former of damp, shaded, or alluvial sites, the latter more frequent in open woods or prairies, both with awnless, few-flowered spikelets.

SIX-WEEKS FESCUE
Festuca octoflora

Importance The value of fescues as livestock forage ranges from poor to excellent. Six-weeks fescue is a warm-season grass which commonly invades overgrazed and burned native ranges and low-fertility tame pastures. The plant characteristically completes its growth in a short time, usually maturing seed in June, then becoming dried. Six-weeks fescue is a poor forage plant for domestic stock. Proper range management and fertilization will control it on both native and improved grasslands.

Fescue

NODDING FESCUE, *Festuca subverticillata*

Nodding fescue, a native cool-season perennial of moist woods, is a good forage plant and readily eaten, especially when green, by all kinds of livestock. However, because it does not comprise a large part of the vegetative cover, it is seldom an important range plant in the Ozarks.

Tall fescue is a cool-season bunch grass used extensively for improved pastures and hay crops. The two strains most frequently used in the Ozarks are Kentucky 31 and Alta fescue. Tall fescue is probably used for reseeding in the Ozark area more than any other tame grass because it is adapted to a wide variety of soils, is aggressive, can withstand heavy utilization better than most native grasses, and responds well to fertilizer. Although frequently used in mixtures with some legumes, tall fescue, if not properly utilized, will crowd out the legumes.

Tall fescue provides green forage with high nutritive content in early spring and late fall when most native grasses are dormant. It is highly palatable and readily eaten by all kinds of livestock. Deer and rabbits will utilize the green foliage in early spring and late fall when other green feed is scarce. Fescue foot, a circulatory deficiency affecting the extremities and causing lameness, may develop in livestock that graze tall fescue exclusively.

The fescues are eaten sparingly by prairie chickens, turkeys, ruffed grouse, and quail.

Hordeum pusillum

Little Barley

Description Low annual of the barley tribe, narrow-tufted, with short, slender culms not exceeding 15–20 in. (35–50 cm.) tall; leaf blades smooth or somewhat scabrous, short, flaglike, to 3/16 in. (5 mm.) wide; sheaths of the upper stem somewhat inflated and often partially enclosing spike; ligule collar-shaped, about 1/16 in. (1½ mm.) long; vernation rolled; spikes narrow, bristly, to 3 in. (7–8 cm.) long, with awned spikelets, the rachis shattering at maturity, flowering in spring.

LITTLE BARLEY, *Hordeum pusillim*

Distribution Widely scattered throughout the region, seldom abundant.

Habitat Fields, open ground, and waste areas; on dry soils.

Foxtail barley, *Hordeum jubatum*, with bushy nodding spikes and long-awned slender spikelets, occurs rarely in the Missouri Ozarks, and is more common in drier ranges to the west.

Importance Little barley is an inferior and undependable forage species. A cool-season grass, it germinates and grows only if moisture is available. After seedheads form in May or June, the sharp spikelets can work into the mouths and around the eyes of animals, causing wounds and infection of the gums, tongue, and eyes. Foxtail barley is a common cause of lumpy jaw in cattle. Horses are even more easily injured as the mucous membranes of

the mouth are thinner and more easily penetrated by the awns and other sharp parts.

Little barley is an invader on abused ranges and pastures. It is a prolific seeder, and once established in an area it is hard to control or eradicate. Good range management practices that result in a vigorous stand of native grasses are the best control.

Rush

Description Low or medium-sized perennials, tufted, or forming colonies, with glabrous stems 12–36 in. (30–90 cm.) tall or more; leaves narrow, stringlike, or absent; flowers small, brownish, in branching clusters from near the summit of the stem, subtended by narrow leaflike bracts, developing in late spring and summer.

PATH RUSH, *Juncus tenuis*

SOFT RUSH, *Juncus effusus*

Distribution Widespread and sometimes common and abundant.
Habitat Fields, roadsides, and open woods, to low ground and alluvial habitats.

Two common species of fields, woods, swales, and low ground, with wiry stems about 12 in. (30 cm.) tall or more are the **path** or **poverty rushes**, *Juncus tenuis*, and *Juncus interior*. The first species is recognized by conspicuous whitish lobes (auricles) at the junction of leaf and stem, the second species by brownish lobes.

Soft rush, *Juncus effusus*, occurs in wet ground and shallow water, forming colonies. The stems are leafless, to 3 ft. (90 cm.) tall or more, unbranched, and tapering to a point with a fascicle of brownish flowers several inches below the tip.
Importance The rushes are generally high in protein and other nutrients, but have only fair forage value. Much of their grazing value for livestock is based on their ability to remain green through most of the year and to start growth early in the season. Their extensive root system enables them to withstand heavy grazing, serves to bind the soil, and helps prevent erosion.

Rushes also provide food for waterfowl and prairie chickens.

Koeleria pyramidata

Junegrass

Description Perennial bunchgrass with culms to 20 in. (50 cm.) tall or more; leaf blades narrow, about ⅛ in. (2–3 mm.) wide, becoming in-rolled when dried; lowermost sheaths pubescent; ligule collar-shaped, minute; vernation folded; panicles narrow, dense, spikelike, with awn-less spikelets, developing in late spring.

JUNEGRASS, *Koeleria pyramidata*

Synonym *Koeleria cristata, Koeleria macrantha*

Distribution Most common in the northern and western parts of the Ozark range.

Habitat Prairies, glades, and open woods.

Importance Junegrass, a cool-season bunchgrass, provides very good to excellent forage for all kinds of livestock; but because it is not very abundant and does not yield a large amount of forage per plant, it is not an important range grass in the Ozarks.

Junegrass usually starts growth and matures seed early in the season. It is considered a decreaser range plant in prairies to the west.

Muhly

Description Native perennials, most of these rhizomatous, solitary or tufted, upright to somewhat decumbent, sparsely or profusely branched, and occasionally forming sprawling mats; ligule collarlike, minute to conspicuous; vernation rolled in the Ozark species; panicles mostly narrow with short closely ascending branches, or broad with spreading branches, the spikelets small, 1-flowered, awned or awnless, developing in middle to late summer.

HAIRGRASS, *Muhlenbergia capillaris*

Distribution Widely scattered throughout the region, sometimes locally abundant.

Habitat Open woods and dry rocky or sandy soils, also moist bottoms, creekbanks, shaded areas, and disturbed ground.

The nonrhizomatous **hairgrass**, *Muhlenbergia capillaris*, of dry soils, is identified by erect tufted habit, narrow leaf blades 1/16–3/16 in. (1–4 mm.) wide, broad showy panicles with fine branching, and spikelets with long straight awns.

Muhly

Muhlenbergia tenuiflora

MEXICAN MUHLY
Muhlenbergia mexicana

NIMBLEWILL, *Muhlenbergia schreberi*

Two woodland types, **Muhlenbergia sobolifera** and **Muhlenbergia tenuiflora**, are distinguishable by rhizomes, erect culms, narrow panicles, and spreading flaglike leaf blades. The first species is typically awnless, the second has a long slender awn.

Species of moist low-lying ground, somewhat decumbent, with top-heavy branching and reclining habit include **wirestem muhly,** *Muhlenbergia frondosa*, and *Muhlenbergia mexicana*. Both species are rhizomatous with spikelike panicles, and both have awned and awnless spikelets.

A common species of shaded areas and disturbed ground is **nimblewill,** *Muhlenbergia schreberi*, with low tufted leafy stems mostly less than 20 in. (50 cm.) tall; panicles slender, delicate, with ascending long-awned spikelets.

Importance These warm-season grasses are only fair forage for livestock. They are usually grazed to some extent when young and green, but become rather unpalatable when mature. They are not important forage plants in the Ozarks because of their rather low nutrient content, relatively small amount of forage per plant, and lack of abundance except in very local situations. Pheasants, quail, and turkeys make light use of nimblewill.

Wheatgrass

Description Perennials of the barley tribe, sod-forming, spreading by rhizomes, with upright culms to about 3 ft. (90 cm.) tall; leaf blades of various widths but less than ⅜ in. (9–10 mm.); auricles prominent; ligule minute, collar-shaped, and fringed on the margin; vernation rolled; spikes stiff, erect, somewhat wheatlike, developing in late spring through the summer.

QUACKGRASS, *Elymus repens*

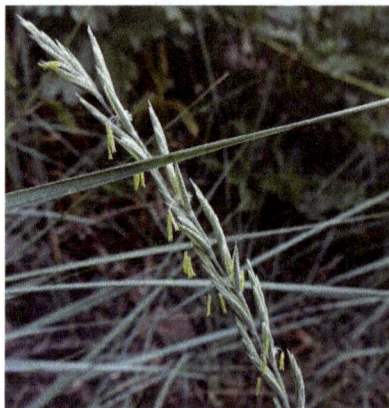

WESTERN WHEATGRASS
Pascopyrum smithii

Distribution Sporadic and generally infrequent, more common northward and westward.

Habitat Prairies, fields, and waste ground; the related common quackgrass, *Elymus repens*, introduced from Europe.

The two most prevalent species are **quackgrass**, formerly treated as *Agropyron repens* (now ***Elymus repens***), with yellowish scaly rhizomes and leaf blades to ⅜ in. (10 mm.) wide; and the native **western wheatgrass**, formerly *Agropyron smithii* (now ***Pascopyrum smithii***), identified by somewhat stiff bluish-green foliage and narrow leaf blades about 1/5 in. (5 mm.) wide or less.

Importance These cool-season grasses are highly palatable and good forage for all classes of livestock. Western wheatgrass and quackgrass, which grow in the fall when moisture conditions are favorable, may remain green during the winter and produce seed in June. These plants are probably grazed by deer in winter and early spring when other green forage is scarce.

Western wheatgrass decreases if closely grazed, especially in spring. Quackgrass and, to a lesser extent, western wheatgrass are aggressive sod-formers and provide good erosion control on embankments and waterways.

Panicum spp., *Dichanthelium* spp.

Panic Grass

Description A large complex group on a worldwide basis, concentrated in warm latitudes, especially in the American tropics, with the greatest number of species for any genus of grasses. In the Ozarks, it is also represented by more species than any other grass genus.

The panic grasses include annuals and perennials of various habits with narrow to wide leaf blades; foliage glabrous to heavily pubescent; ligule mostly tufted or consisting of a fringe of hairs; vernation rolled; panicles slender or broad with spreading branches, sometimes from axils of sheaths, 1-flowered and awnless, developing in spring or summer, or in both spring and summer in some species.

Dichanthelium latifolium

Distribution Widespread, or some species with limited range.

Habitat On various sites, including glades, prairie openings, forest areas, waste ground; on dry cherty soils to moist alluvium.

Important species are grouped according to the following characteristics: (1) Tufted branching perennials with winter rosettes and spring and fall panicles, the latter from the leaf sheaths, (2) erect mostly coarse perennials, flowering once, and (3) upright or decumbent tufted annuals.

The first group is prevalent in the Ozarks and includes the following common species, now placed in genus *Dichanthelium*: ***Dichanthelium linearifolium*** (synonym *Panicum linearifolium*), with slender tufts of narrow ascending leaf blades usually less than 3/16 in. (4–5 mm.) wide and 20 times longer than broad, of dry upland woods

and open ground; ***Dichanthelium boscii*** (synonym *Panicum boscii*), and ***Dichanthelium latifolium*** (synonym *Panicum latifolium*) of woods habitats, with strikingly wide leaves to 1 in. (2½ cm.) or more, and usually not over 5 times as long; ***Dichanthelium commutatum*** (synonym *Panicum commutatum*), ***Dichanthelium dichotomum*** (synonym *Panicum dichotomum*), and ***Dichanthelium acuminatum*** (synonym *Panicum lanuginosum*), representing an intermediate leaf type with leaf blades 5 to 20 times longer than wide, these species typical of woods, fields, and waste places.

Panic Grass

SWITCHGRASS, *Panicum virgatum*

WITCHGRASS, *Panicum capillare*

FALL PANIC GRASS
Panicum dichotomiflorum

The second group includes the **switchgrass**, *Panicum virgatum*, of prairies and glades, forming clumps or stands, spreading by rhizomes, with leafy flower stalks to 5 ft. (1.5 m.) or more; leaf blades glabrous but with tuft of hairs at junction with sheath; panicles terminal, somewhat stiff with spreading branches and pointed spikelets, flowering in summer.

In the third group, three warm-season annual species are common. **Witchgrass**, *Panicum capillare*, is a tufted decumbent-branching species, with hispid-pubescent sheaths and broad terminal panicles. *Panicum flexile* has a more erect slender habit and narrower panicles. Both of these species are found in dry fields, waste areas, glades, and open woods. A third annual, **fall panic grass**, *Panicum dichotomiflorum*, is easily distinguished by tufted coarse decumbent stems with smooth sheaths and leaf blades. It is prevalent in waste areas and on low or damp ground.

Panicum spp., *Dicanthelium* spp.

Panic Grass

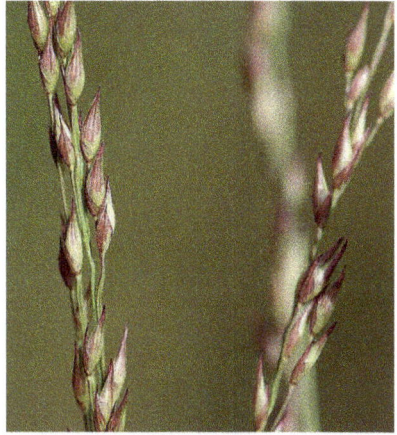

SWITCHGRASS, *Panicum virgatum*

Importance The seeds and foliage of panic grasses are important foods for several wildlife species. Basal rosettes of the first group of species grow during autumn, remain green through winter, and provide valuable winter and early spring forage for deer. Turkeys utilize these rosettes plus foliage and seeds of other species of *Panicum* throughout the year. Ducks and geese make substantial use of fall panic grass (*P. dichotomiflorum*), a heavy-seed-producing annual, and light use of other species. Fall panic grass and, to a lesser extent, witchgrass provide food for doves and quail. Ruffed grouse, pheasants, and prairie chickens eat minor amounts of these plants.

As livestock forage, the panic grasses range from worthless to good or, some, very good. Two warm-season annuals, *Panicum flexile* and *P. capillare*, are poor to fair forage when young but mostly unpalatable when mature. They are generally considered invaders in overused native ranges and pastures. *P. commutatum*, *P. dichotomum*, and *P. boscii*, are warm-season shade-tolerant perennials common in forest ranges throughout the Ozarks. They provide fair forage when green but are unpalatable when dried and mature. These species do not produce much forage per plant and are not very abundant, but will increase with overgrazing as the more desirable forage grasses decrease.

Switchgrass is the most important panic grass in the Ozarks for livestock forage. This warm-season grass is eaten by all kinds of livestock and provides nutritious green forage or prairie hay. It begins growth in April and usually matures seedstalks earlier than other prairie grasses.

Switchgrass is grazed principally in the spring and early summer before the leaves become coarse and tough. It is a very heavy forage producer and has been used in seeding native grass pastures for hay, either by itself or in mixtures. In pure stands, it can produce more than 2 tons of wild hay per acre. Continued overgrazing during the growing season causes switchgrass to be replaced by less productive species.

Paspalum

Description Perennials in our range, solitary or tufted, some species with short rhizomes; culms spreading to erect, to 4½ ft. (1½ m.) tall but usually shorter; foliage glabrous to heavily pubescent; ligule collar-shaped; vernation rolled; inflorescence consisting of short racemes, the spikelets orbicular to oblong, semispherical and arranged on one side of the winged rachis, developing in middle to late summer.

FRINGE-LEAF PASPALUM
Paspalum setaceum

Paspalum laeve

Distribution Widely scattered throughout the region, seldom abundant.

Habitat Prairies, glades, open woods, roadsides, and waste ground; on dry to damp soil; some species introduced from Central and South America.

Two common species are **Fringe-leaf paspalum**, *Paspalum setaceum*, and *Paspalum laeve*. The first species has spikelets about 1/16 in. (1½ mm.) long or slightly more, and occurs in dry habitats; the latter species has larger spikelets ⅛ in. (3 mm.) long, and usually occurs in more moist sites.

Two less prevalent species, mostly robust, occupying damp swales, roadsides, and borders of woods are *Paspalum pubiflorum* with spikelets in four rows on the rachis and *Paspalum floridanum*, a coarse grass with stiff racemes and two rows of large spikelets to 3/16 in. (4 mm.) long.

Paspalum

Paspalum pubiflorum

The well-known **bahiagrass**, *Paspalum notatum*, and **dallisgrass**, *Paspalum dilatatum*, are forage grasses in the southern United States, introduced from Central and South America.

Importance Paspalums generally begin growth in early spring and mature seed from early to late summer.

Fringeleaf paspalum provides important summer and early fall food for turkeys. It is also eaten by doves, quail, prairie chickens, and raccoons. Other paspalums are used only lightly.

These warm-season grasses are not important to livestock in the Ozarks because they are not abundant and do not produce much forage per plant. The paspalums are grazed mostly in the spring and early summer when green and succulent.

BAHIAGRASS
Paspalum notatum

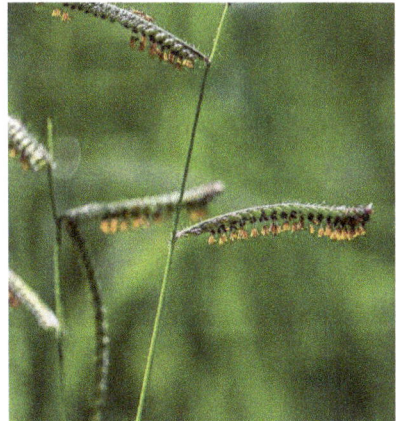

DALLISGRASS
Paspalum dilatatum

Timothy

Description Tall leafy perennial, forming clumps; culms with bulbous bases, to about 40 in. (1 m.) tall; leaves flat, elongate, up to ⅜ in. (10 mm.) wide; ligule conspicuous, membranous, about ⅛ in. (2–3 mm.) long; vernation rolled; flowering heads cylindrical, spikelike to about 6 in. (15 cm.) long, developing in late spring and early summer.

TIMOTHY, *Phleum pratense*

Distribution Widespread and sometimes common and abundant.
Habitat Meadows, fields, roadsides, and waste ground; introduced from Europe and widely naturalized.
Importance Timothy is one of the most important perennial cultivated plants in North America. It is widely used for hay and pasture.

This cool-season short-lived grass is moderately abundant in the Ozarks, but is probably best adapted to cooler humid habitats further north. It begins growth early in the season and the seeds are usually formed by midsummer.

Timothy is highly nutritious and provides good forage for cattle, horses, and sheep. It also makes up part of the unidentified grass eaten by deer and other wildlife in early spring; seeds are eaten by doves.

Timothy maintains itself well if properly used, but is not resistant to heavy grazing as the basal bulbs are easily injured by close cropping and heavy trampling. It responds well to fertilizer and produces a large amount of forage per plant.

Timothy makes a highly palatable hay valued for horses, but is inferior to alfalfa and clover hays for fattening cattle.

Poa spp.

Bluegrass

Description Annuals and perennials; tufted, or sod-forming and spreading by rhizomes, low, decumbent to erect; leaf blades mostly keeled with boat-shaped tip; ligule collar-shaped; vernation folded; panicles narrow or with spreading branches; spikelets small, several-flowered, most species with cobwebby pubescence at base of floret, awnless, developing in spring or early summer.

KENTUCKY BLUEGRASS, *Poa pratensis*

Distribution Widespread and sometimes common and abundant.
Habitat Fields, roadsides, pastures, open woodland, and moist shaded habitats; *Poa pratensis*, *Poa compressa*, and *Poa annua* are introduced from Europe.

A common species is **Kentucky bluegrass**, *Poa pratensis*, forming dense sod and spreading by rhizomes; foliage glabrous, the leaf blades about ⅛ in. (2–4 mm.) wide, with distinct boatshaped tip; ligule whitish, small, about 1/32 in. (½ mm.) long; panicles erect, pyramidal with spreading whorled branching.

Canada bluegrass, *Poa compressa*, occurs commonly on poorer sites. It differs from Kentucky bluegrass by its flattened stems or culms and narrower panicles with short ascending branches, and forms a less dense sod.

Two native perennials of shaded sites are *Poa sylvestris* and *Poa wolfii*. These form loose sparse tufts and lack rhizomes.

Annual bluegrass, *Poa annua*, abundant in spring in open ground and waste areas, is identified by its low tufted habit, flat leaf blades, and individual florets that lack the cobwebby pubescence of other *Poa* species.

Bluegrass

ANNUAL BLUEGRASS, *Poa annua*

Importance Kentucky bluegrass and Canada bluegrass, both cool-season long-lived grasses, are among the most important forage grasses in the United States. However, they are not as abundant or productive on Ozark native ranges with their high summer temperatures and the relatively infertile and drouthy soils as in cooler northern regions. Both species usually start growth in late winter, produce seed by June, become dormant during the summer, and resume growth in the fall.

The bluegrasses are highly palatable to all kinds of livestock. They are rated as very good forage for cattle and horses, and good for sheep. Bluegrasses are generally heavily grazed in early spring when they are green and succulent and when the native warm-season grasses have not yet started growth, and again in the fall after most of the native grasses have dried.

These grasses are shade tolerant and are commonly found in wooded pastures, where they may remain green all summer. They are very susceptible to early spring fire, and will eventually disappear under continued annual spring burning.

The bluegrasses are good seed producers, aggressive sod formers. and well-suited for erosion control. Kentucky bluegrass is also widely used in lawns.

Scirpus spp., *Schoenoplectus* spp.

Bulrush

Description Low to tall perennials of the sedge family, to 3–4 ft. (90–120 cm.) tall or more; stems solid or pithy, with closed sheaths, mostly 3-angled or sometimes round in cross section.

GREAT BULRUSH
Schoenoplectus tabernaemontani

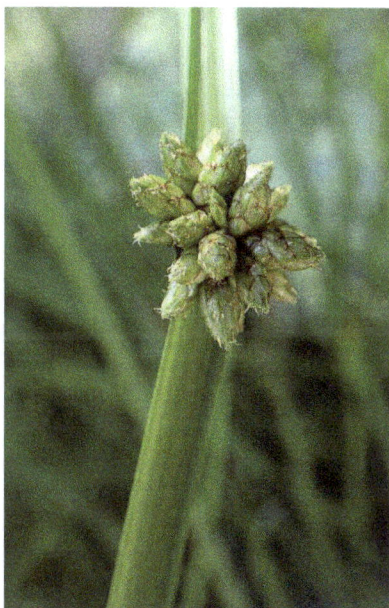

THREE-SQUARE BULRUSH
Schoenoplectus americanus

Distribution Widely scattered throughout the region, sometimes locally abundant.

Habitat Wet or damp ground, sloughs, ponds, seeps, drainages, along streams and gravel bars.

A common species is **great bulrush**, *Schoenoplectus tabernaemontani* (synonym *Scirpus validus*), usually more than 3 ft. (90 cm.) tall, round in cross section, dark green, with a cluster of brownish scalelike flowers near the summit. It is leafless or has leaves only at the base of the stem. **Three-square bulrush**,

Schoenoplectus americanus (synonym *Scirpus americanus*), is shorter, sharply triangular, and leafless, with a conelike cluster of scaly flowers below the tip. Both species form colonies in shallow water.

A bulrush with leafy stems, usually tufted, growing in wet ground, sloughs, low prairies, and glade seeps is **dark green bulrush**, *Scirpus atrovirens*, with terminal branching flower clusters, upright pedicels, and dark brown spikelets at maturity.

Bulrush

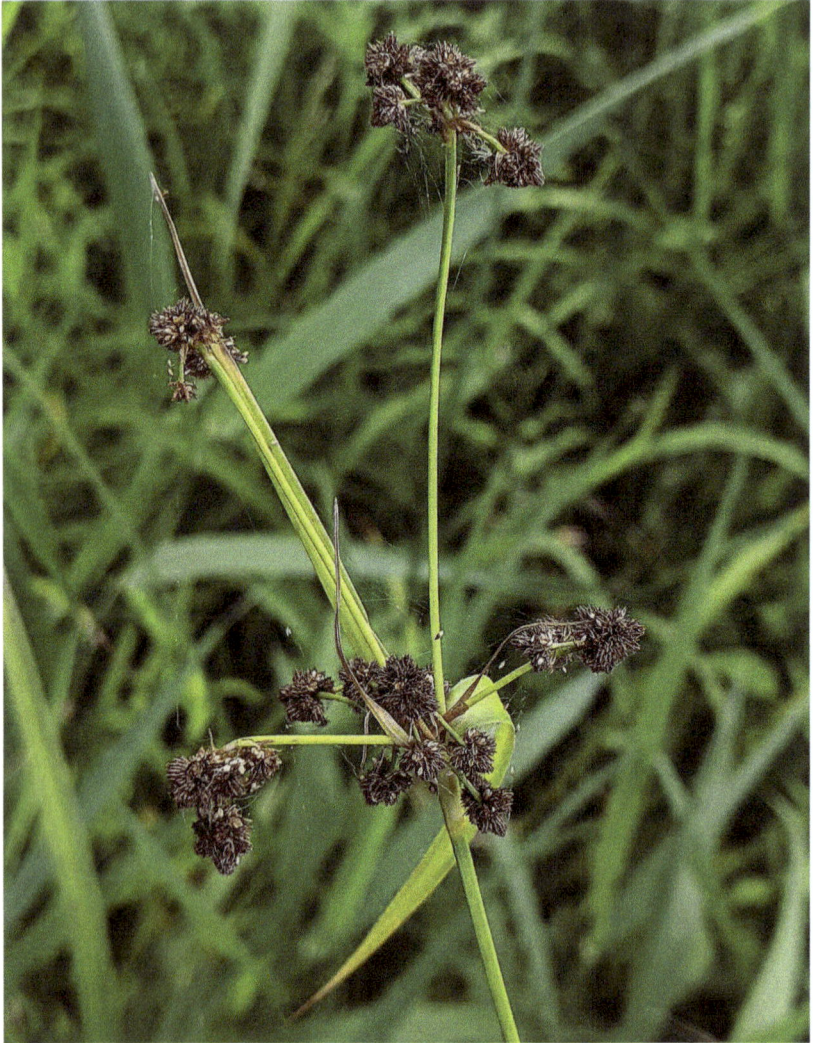

DARK GREEN BULRUSH, *Scirpus atrovirens*

Importance The "seeds" or achenes of bulrushes are an important food for ducks. Geese and pheasants also utilize bulrushes but to a lesser extent.

Bulrushes rate as poor to fair forage for livestock. However, they provide some forage during critical times of the year since they begin growth early in the spring and remain green until late in the season.

Foxtail, Bristlegrass

Description Species included here are tufted annuals of variable height, erect or decumbent with branching near base; sheaths overlapping; ligules consisting mostly of minute hairs; vernation rolled although sheath somewhat keeled; inflorescence cylindrical, bristly, spikelike with plump grainlike spikelets, developing from midsummer to fall.

GREEN FOXTAIL, *Setaria viridis*

YELLOW FOXTAIL, *Setaria pumila*

Distribution Widespread and sometimes common and abundant.
Habitat Fields, waste areas, and disturbed ground; the species described here are introduced from Europe or Asia.

Two common species are **green foxtail**, *Setaria viridis*, and **yellow foxtail**, *Setaria pumila*, both mostly glabrous or with ciliate hairs on sheath and leaf margins. Green foxtail is identified by somewhat nodding heads, greenish or purplish, about 2 in. (5 cm.) long, and yellow foxtail by erect dense heads,

yellowish, 2–4 in. (5–10 cm.) long.

An adventive becoming more widespread in recent years is **nodding foxtail**, *Setaria faberii*, a coarse species to 4½ ft. (1½ m.) tall or more, with pubescent foliage and nodding panicles 4 in. (10 cm.) long, resembling *Setaria viridis* but a much larger plant.

The cultivated **foxtail millet** or **Italian millet**, *Setaria italica*, with thick lobulate heads 1–1½ in. (2½–4 cm.) wide and as much as 10 in. (25 cm.) long is an occasional volunteer in open ground.

Foxtail, Bristlegrass

NODDING FOXTAIL, *Setaria faberii*

FOXTAIL MILLET, *Setaria italica*

Importance Foxtails are probably more important than any other wild grass in the Ozarks as food for quail and doves (only corn and wheat are more important). Foxtails also provide minor foods for turkeys, deer, and geese and are occasionally eaten by grouse and raccoons.

These warm-season grasses are moderately palatable to most kinds of livestock. They usually begin growth in April and develop seedstalks from June to September. They are grazed primarily in the spring and early summer when green and succulent and are seldom eaten when the plants are dried or mature. The foxtails do not produce a great amount of forage per plant and are not dependable forage producers.

Some foxtail millet was used at one time for pasture, hay, and silage, but has been largely replaced for these purposes by more productive grasses. The foxtails now are considered weeds in cultivated fields and invaders in native rangelands.

Indian grass

Description Tall coarse perennial, to 7½ ft. (2½ m.), forming clumps, from short rhizomes; leaf blades and sheaths mostly glabrous to slightly pilose; ligule conspicuous, stiff, 3/16 in. (4–5 mm.) long; vernation rolled; panicles plumelike, shiny brown, with numerous short ascending branches, the spikelets dark brown, hirsute, long-awned, developing in late summer. This species is a most attractive grass when in full flower.

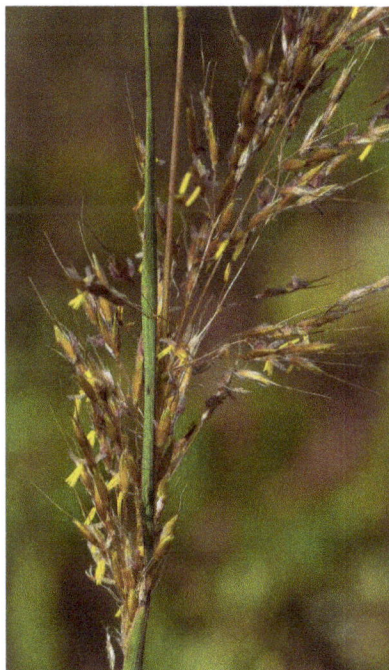

INDIAN GRASS, *Sorghastrum nutans*

Distribution Widespread and sometimes common.

Habitat Prairies, glades, open woods, and clearings.

Importance This warm-season grass is an excellent forage plant and is one of the most important native tall grasses. It is highly nutritious and readily eaten by all kinds of livestock, especially when the plants are young and green. This grass provides a large amount of forage and makes excellent hay. It can be established from seed and will produce 2–3 tons of cured hay per acre. It is becoming increasingly popular for seeding as a hay grass, either by itself or in mixtures with other native grasses.

Continuous close grazing before maturity reduces plant vigor and will ultimately result in replacement of Indian grass by less desirable species.

Sudan grass, Johnson grass

Description Tall coarse annuals or perennials, the latter with thick creeping rhizomes; culms to 8–9 ft. (2½–3 m.) tall; leaf blades of varying width; ligule collarlike, conspicuous; vernation rolled; panicles large, terminal, with spreading branches and spikelets in pairs, one sessile, awned, the other stälked, awnless, developing in summer.

Distribution Occurs sporadically throughout the region, sometimes locally abundant and forming dense stands.

Habitat Fields and waste ground; the species from the Old World introduced as forage plants and frequently volunteering or spreading.

The well-known **sudan grass**, **Sorghum x drummondii**, is a tall leafy annual with mostly glabrous leaves to ⅝ in. (15 mm.) wide or more. Sorghum, kafir, milo, and broomcorn are other common crop plants classified in the genus Sorghum.

Johnson grass, *Sorghum halepense*, is a tall leafy perennial and a vigorous spreader, distinguished from sudan grass by coarse creeping rhizomes; leaf blades glabrous, ⅓–1 in. (8–25 mm.) wide, with whitish midrib and occasionally splotched with purple.

Importance The grain of several species of Sorghum rank as important food for quail and doves. Deer feed on the grain, especially during fall. Raccoons also use these plants for food. Quail, geese, and turkeys make light use of Johnson grass.

These warm-season grasses are moderately palatable for livestock and considered fair forage. They all

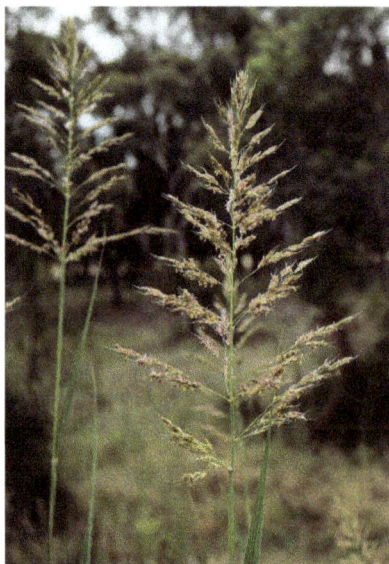

JOHNSON GRASS
Sorghum halepense

produce a large amount of growth per plant and are extensively used for pasture and hay.

However, Johnson grass can completely take over a site to the detriment of better forage grasses. It is difficult to control because of its extensive rhizomes and aggressive and persistent growth habit.

Caution must be exercised in using these grasses as forage. Livestock occasionally have been killed by prussic acid (hydrocyanic acid) poisoning from grazing these plants during dry summers or immediately after frost.

Sphenopholis spp.

Wedgegrass

Description Perennials forming loose narrow tufts; culms erect, 2–3 ft. (60–90 cm.) tall; leaf blades narrow ⅛– 3/16 in. (3–4 mm.) wide or less; ligule whitish, membranous, lacerated on the margin; vernation rolled; panicles spikelike, with short ascending branches, developing in spring and summer.

ligule

Distribution Widely scattered throughout the region, generally not abundant.

Habitat Dry open woods, glades, and prairies to damp shaded habitats.

Prairie wedgegrass, *Sphenopholis obtusata* has compact panicles and short lobelike branches, and occurs in open woods and glades. *Sphenopholis intermedia* of damp or shaded ground is identified by less dense flower stalks and longer panicle branches. Junegrass, *Koeleria pyramidata*, somewhat similar to *S. obtusata*, is distinguished by its folded vernation, collar-shaped ligule, and narrower leaf blades.

Importance These native warm-season grasses are moderately palatable and are fair forage for most kinds of livestock. They are grazed mostly when young and green, and seldom eaten when the

PRAIRIE WEDGEGRASS
Sphenopholis obtusata

plants become dried and mature. These grasses are not abundant and are not considered important forage plants in the Ozarks. However, they provide a source of food for turkeys primarily in late spring and early summer.

Dropseed

Description Annuals and perennials, most species either tufted, erect, stout, or stems weakly decumbent and spreading; leaf blades narrow, curling or inrolled, usually not more than ⅛ in. (2–3 mm.) wide; ligule an inconspicuous fringe of minute hairs; vernation rolled; panicles narrow, spikelike, partially inserted in sheath, or diffuse, exserted, with spreading branches, the spikelets 1-flowered and awnless, developing in summer.

Distribution Widespread and sometimes common and abundant.
Habitat Open woods, prairies, glades, roadsides, and waste ground; primarily on dry soils.

Two common annuals of glades, open woods, and waste areas are **baldgrasses**, *Sporobolus neglectus* and *Sporobolus vaginiflorus*, identified by low tufted habit and somewhat decumbent stems, generally less than 18–30 in. (50 cm.) tall; leaf blades narrow, inrolling; sheaths inflated, and partially or wholly including the narrow spikelike panicles. These annuals are most reliably separated on the basis of spikelet characters, the first species usually plump, glabrous, and the latter narrow and somewhat pubescent.

Tall dropseed, *Sporobolus compositus* (synonym *Sporobolus asper*), is a robust perennial of glades, dry woods, and waste ground, primarily in the northern and western counties, erect, to 3 ft. (90 cm.) tall or more, with stringlike leaf tips and expanded sheaths enveloping the narrow panicles. *Sporobolus clandestinus*, occurring throughout the Ozarks, is similar but is generally

BALDGRASS, *Sporobolus vaginiflorus*

shorter and less robust, and should be separated with certainty on the basis of spikelet characters. The spikelets in *S. compositus* are somewhat plump, with glabrous lemmas; those in *S. clandestinus* are elongate-pointed and slightly pubescent.

Dropseed

TALL DROPSEED
Sporobolus compositus

PRAIRIE DROPSEED
Sporobolus heterolepis

Prairie dropseed, *Sporobolus heterolepis*, is a tufted perennial, with dense basal foliage of fine, stringlike leaves and, unlike all the preceding species, has open panicles with spreading branches, developing a sweetish odor during flowering.

Importance Tall dropseed and prairie dropseed are warm-season grasses. They are not abundant and never make up a large part of the vegetative cover, but are found generally throughout the Ozarks. Tall dropseed is moderately palatable to livestock and is considered fair to good forage. It is grazed when young and green and seldom eaten when dried or mature.

Prairie dropseed is highly palatable to all kinds of livestock and is considered good forage. It is grazed readily when young and green, and also furnishes fair winter forage.

Annual dropseed, also known as baldgrass, is palatable when green. However, it loses nutritive value very rapidly upon curing, is not readily grazed in late summer, and does not produce much forage per plant. It is an invader on abused ranges, and may form nearly pure stands.

The seeds of baldgrass are eaten in measurable amounts by quail, and in smaller amounts by prairie chickens. Turkeys eat seeds of tall dropseed in winter.

Purpletop, Grease Grass

Description Tall, somewhat coarse perennial, forming leafy clumps, with flower stalks to about 4½ ft. (1.3 m.) tall; leaf blades and sheaths mostly glabrous; sheaths strongly flattened at base of plant, and with tuft of hairs at junction with leaf blade; ligule minute, fringed; vernation rolled; panicles with spreading or drooping branches, purplish, somewhat sticky or viscid, developing in mid- to late-summer.

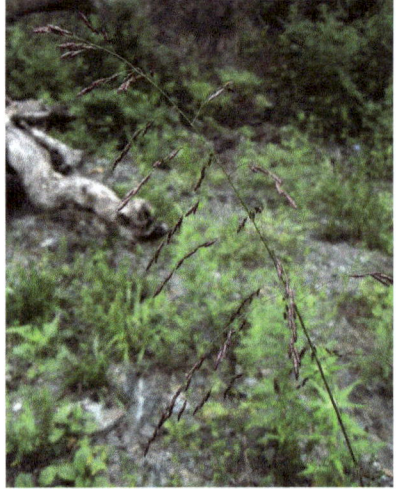

PURPLETOP, *Tridens flavus*

Distribution Widespread and sometimes locally abundant.

Habitat A common species of roadsides, waste ground, open woods, and glades.

Importance This grass begins growth in early spring and develops seedstalks from July to September. It is moderately palatable and readily eaten by all kinds of livestock in the early stages of growth but becomes coarse and tough when mature and is seldom eaten at this stage.

On native ranges, purpletop will increase as the better grasses decrease by heavy use. If continuously grazed closely, it will also be eliminated and replaced by less productive plants.

Purpletop is an important source of food for turkeys in late summer. Quail make slight use of the seeds, and deer occasionally eat the plant.

Tripsacum dactyloides

Gamagrass

Description Tall coarse perennial, forming large leafy clumps from short thick rhizomes; flower stalks to 6–7 ft. (2 m.) tall; leaf blades mostly glabrous, about ⅜–1 in. (1–2½ cm.) wide; ligule a fringe of short hairs; vernation rolled; flower stalks with several digitate or fingerlike spikes, staminate spikelets in upper part and pistillate spikelets below on each spike, the latter spikelets recessed in the thick hard rachis joint, developing in late spring and summer.

GAMAGRASS, *Tripsacum dactyloides*

Distribution Widely scattered throughout the region, seldom abundant.
Habitat Prairies, swales, low ground, and waste areas; most generally on deep or moist soils.
Importance Gamagrass is a warm-season grass and a high forage producer, but not sufficiently abundant to be an important plant in the Ozarks. Because it is highly palatable to livestock, it is readily grazed out in native pastures.

Keys to the Genera

HOW TO USE THE KEYS

The plant keys presented here are based primarily on vegetative characters, which are generally easier to discern and persist longer than do the flowers and fruit. The keys identify most plants only to the genus level. Additional discussion by species is provided in the text.

A small ruler to measure parts of specimen plants and an 8- to 10-power hand lens will be useful. The reader untrained in botany may refer to the glossary for definitions of plant terms.

At each step in the key, contrasting statements describing plant characteristics are given. Select the one that most closely describes the unknown plant and note the number on the right margin of the page that indicates the next pair of alternatives to be evaluated. Proceed to that number pair and again select the more appropriate choice. Continue in this manner until the plant is identified. For example, in step 1 you decide that your sample is herbaceous rather than woody and proceed to step 63. Absence of flowers or seeds and the leafless stems leads to step 64. Tubular stems and whorls of naked branches rather than leaflike fronds indicate that your specimen is *Equisetum*. Reference to the text will help verify your determination.

The keys were designed to identify only that segment of the total flora of the Ozark region that is included in this book. Because of general similarities of plant characters, it is possible to identify from the keys a plant not included in the book. To decide whether the identification is correct, always refer to the followup descriptions and photographs for plants of the genus.

SUMMER KEY

1   Plants woody, all trees, many shrubs and vines: or appearing semi-woody as in *Acaciella angustissima* and *Amorpha canescens*, both low shrubs with numerous small leaflets 2

1   Plants herbaceous; annuals or perennials, including both grasses and nongrass species 63

2   Leaves evergreen, either scalelike, less than ¼ in. (6–7 mm.) long, or needlelike, several inches long 3

2   Leaves deciduous (falling at end of growing season), not scalelike or needlelike as above 4

3   Leaves small, scalelike; evergreen trees of limestone glades, old fields, pastures, and waste ground. . . . . **Cedar (*Juniperus* spp.), p. 44**

3 Leaves needlelike, in bundles of 2 or 3; native pine tree of dry slopes and ridges, on acid soils .... **Shortleaf pine (*Pinus echinata*),** p. 55

4 Leaves simple, not divided into several distinct leaflets 5
4 Leaves compound, with 3 or more distinct leaflets 48

5 Leaves opposite each other on the stem. 6
5 Leaves alternate, 1 at each node 16

6 Leaves with smooth margins, neither serrate on the edge of blade nor lobed (sometimes lobed in Japanese honeysuckle) 7
6 Leaves with serrate margins, or lobed 12

7 Leaves ovate to elliptic; flowers tubular, showy, red, orange, or cream colored; plants bushy with arching stems, or trailing and vinelike .....
.... **Honeysuckle (*Lonicera* spp.),** p. 47
7 Leaves ovate, broadly elliptic, oblong, oblong-lanceolate, or obovate; shrubs or small trees... 8

8 Leaves ovate to broadly elliptic with pointed apex and conspicuous arching veins; flowers cream colored, greenish white, or white; fruit berrylike, white, blue, or red; shrubs or small understory trees (leaves alternate in *Cornus alternifolia*) .... **Dogwood (*Cornus* spp.),** p 28.
8 Leaves rounded, oblong, oblong-lanceolate, or obovate; shrubs 9

9 Leaves ovate or rounded, less than 2 in. (5 cm.) long; fruit small reddish berries clustered in the axils of leaves; low shrub of fields, pastures, and open woods.......... **Coralberry (*Symphoricarpos orbiculatus*),** p. 78
9 Leaves broadly elliptic, oblong-lanceolate, obovate, or spatulate..... 10

10 Leaves broadly elliptic, tapering to both apex and base, peddled, often 3 at a node; flowers small, in spherical clusters about 1 in. (2½ cm.) or more in diameter; tall shrub usually 6–10 ft. (2–3 m.), on wet ground or near water .... **Buttonbush (*Cephalanthus occidentalis*),** p. 26
10 Leaves oblong or obovate to spatulate, tapering toward base, without petiole.. 11

11 Leaves small, oblong to obovate, about ¾ in. (2 cm.) long or less; flowers pale yellow with 4 petals; stems usually several from base, branching, with reddish scaly bark, about 12–15 in. (30–40 cm.) tall ....
.... **St. Andrew's Cross (Ascyrum hypericoides),** p. 42
11 Leaves oblong to spatulate, with translucent dots in some plants; flowers yellow with 5 petals; shrubs to 7 ft. (2 m.) ....
.... **St. John's-wort (*Hypericum* spp.),** p. 40

12 Leaves lobed, without serrate margins; fruit a double-winged samara; forest trees .... **Maple (*Acer* spp.),** p. 13
12 Leaves not lobed, with serrate margins; shrubs or understory trees .. 13

13 Bark of young twigs smooth, green; flowers small, purplish; fruit splitting, with bright red "seeds" **Eastern wahoo (*Euonymus atropurpureus*),** p. 34
13 Bark sometimes scurfy or hairy, brown or reddish; flowers white or pinkish blue; fruit not as above 14

14   Bark scaly or stringy; flowers in flat open clusters, white; shrub of moist woods or bluffs . . . . **Wild hydrangea (*Hydrangea arborescens*)**, p. 39

14   Bark not stringy; flowers in dense clusters, white or pinkish blue; understory trees or shrubs 15

15   Leaves and twigs hairy; flowers pinkish, in small clusters at base of leaves; fruit lavender, sometimes white . . . . **French mulberry (*Callicarpa americana*)**, p. 19

15   Leaves and twigs smooth or sometimes scurfy; flowers white; fruit black, with single flat seed . . . . **Blackhaw (*Viburnum* spp.)**, p. 84

16   Leaves with smooth or wavy margins 17

16   Leaves with serrate margins, or lobed 26

17   Branches thorny or spiny 18

17   Branches not thorny or spiny 20

18   Plants shrubby, vinelike, climbing or twining; leaves varying from broadly ovate to cordate to arrow-shaped . . . . . . **Greenbrier (*Smilax* spp.)**, p. 76

18   Small to medium-sized trees 19

19   Leaves ovate, long-pointed, rounded at base . . . . **Osage-orange (*Maclura pomifera*)**, p. 48

19   Leaves elliptic to obovate, narrowed at base, with blunt tips . . . . **Gum bumelia (*Sideroxylon lanuginosum*)**, p. 75

20   Leaves cordate or heart-shaped; pods 2–3 in. (5–7 cm.) long; small tree . . . . **Eastern redbud (*Cercis canadensis*)**, p. 27

20   Leaves not cordate 21

21   Leaves at least 1½–2 in. (4–5 cm.) wide 22

21   Leaves narrower 25

22   Leaves large, occasionally 10 in. (25 cm.) long or more, tapering to base; leaf buds and young twigs with rusty pubescence; flowers dark purplish, about 1 in. (2½ cm.) wide; small understory tree of moist shaded habitats . . . . **Pawpaw (*Asimina triloba*)**, p. 17

22   Leaves smaller; flowers greenish, yellowish white, or bluish gray 23

23   Leaves obovate or widest above the middle, turning scarlet in fall; forest tree on cherty or acid soils . . . . **Blackgum (*Nyssa sylvatica*)**, p. 52

23   Leaves rounded-elliptic to oblong-lanceolate 24

24   Leaves rounded-elliptic, glaucous or bluish green; common tree of limestone glades . . . . **American smoketree (*Cotinus obovatus*)**, p. 31

24   Leaves oblong-lanceolate, not glaucous; common tree of old fields, but occurring also in glades and bottom lands . . . . **Persimmon (*Diospyros virginiana*)**, p. 33

25   Leaves elliptic-oblong, bluish green, with distinct parallel veins from midrib to edge of blade; semi-vinelike shrub in limestone glades, or tall climbing vine in wet woods and lowlands . . . . **Supplejack (*Berchemia scandens*)**, p. 18

37   Leaves variously shaped; staminate catkins pendulous; fruit an acorn .
     . . . . **Oak (*Quercus* spp.), p. 58**

38   Leaves ovate-cordate, with irregular margins; buds obtuse or rounded;
     shrub . . . . **Hazelnut (*Corylus americana*), p. 30**

38   Leaves ovate-oblong, margins mostly even or regular; buds narrow-
     pointed; small tree . . . **Eastern hophornbeam (*Ostrya virginiana*), p. 53**

39   Leaves with 5 pointed lobes, long-stalked; spherical, about 1 in. (2½ cm.)
     in diameter . . . . **Sweetgum (*Liquidambar styraciflua*), p.**

39   Leaves not lobed **40**

40   Leaves ovate or elliptic; flowers small, white, in dense sprays; fruit a dry
     capsule; small shrub. . . . . **New Jersey tea (*Ceanothus* spp.), p. 23**

40   Leaves ovate, cordate, or oblong-lanceolate; flowers not as above . . . **41**

41   Leaves ovate-cordate with pointed tip; fruit dry, berrylike, from an oblong
     leafy bract; forest tree . . . . **American basswood (*Tilia americana*), p. 79**

41   Leaves oblong-lanceolate or, if ovate, not cordate; fruit not as above **42**

42   Leaves coarsely serrate or with irregular or wavy margin **43**

42   Leaves mostly with fine serrations along even margin **44**

43   Leaves mostly oval, wavy margined; flowers yellowish to reddish with
     small stringlike petals appearing in fall and winter or early spring; shrub
     . . . . **Ozark witch-hazel (*Hamamelis vernalis*), p. 38**

43   Leaves oblong-lanceolate, coarsely serrate; flowers large, showy, mostly
     pink; small tree, sometimes spurlike twigs present . . . .
     . . . . **Wild crab (*Malus ioensis*), p. 49**

44   Flowers conspicuous, white, appearing before or with leaves **45**

44   Flowers inconspicuous, small, greenish **46**

45   Leaves oblong-obovate, with whitish hairs when young; buds elongate,
     tapering to narrow point, up to ¼ in. (6–7 mm.) long; fruit berrylike, be-
     coming dry, with 10 seeds . . . .
     . . . . **Downy serviceberry (*Amelanchier arborea*), p. 16**

45   Leaves ovate-lanceolate to obovate, not hairy; buds blunt, shorter than
     above; fruit fleshy with single rounded pit . . . .
     . . . . **Wild cherry (*Prunus* spp.), p. 56**

46   Leaves oblong-lanceolate, with symmetrical base and minute serrations;
     fruit black, berrylike; small tree . . . .
     . . . . **Carolina buckthorn (*Rhamnus caroliniana*), p. 62**

46   Leaves ovate to obovate, usually with oblique base, toothed margins **47**

47   Leaves with veins more or less equally spaced and parallel from midrib;
     fruit orbicular, winged . . . . **Elm (*Ulmus* spp.), p. 80**

47   Leaves with veins not as above, main veins from base of lf.; fruit berrylike,
     orange to purplish with thin flesh and large seed; twigs with "witches
     brooms" . . . . **Hackberry (*Celtis* spp.), p. 25**

48   Branches or stems prickly or thorny **49**

59 Leaves pinnately compound; fruit winged. 60

60 Leaflets 3–5, with few coarse irregular serrations; fruit a double-winged samara . . . . **Boxelder** (*Acer negundo*), p. 13

60 Leaflets mostly 7–9, wavy edged to smooth or finely serrate; fruit a single-winged samara . . . . **Ash** (*Fraxinus* spp.), p. 35

61 Leaves large, doubly compound with numerous egg-shaped, smooth-edged leaflets; fruit a thick heavy pod 5–8 in. (12–20 cm.) long with seeds about ½ in. (1.2 cm.) wide . . . . **Kentucky coffeetree** (*Gymnocladus dioicus*), p. 37

61 Leaves singly compound; leaflets. 5–25 with serrate margins; fruit a nut with thick hull 62

62 Terminal leaflets generally larger than lateral leaflets., if about the same size, the buds yellowish; twigs with solid pith**Hickory** (*Carya* spp.), p. 20

62 Terminal leaflets smaller than lateral leaflets; buds not yellowish; twig with brownish chambered pith . **Butternut, walnut** (*Juglans* spp.), p. 43

63 Plants producing spores; flowers and seeds not present; fernlike or with leafless stems 64

63 Plants producing flowers and seeds 66

64 Plants leafless, with tubular stems, sometimes with whorls of naked branches . . . . **Horsetail** (*Equisetum* spp.), p. 128

64 Plants with leaflike fronds; native ferns. 65

65 Plants low-spreading, nonbranching; fronds elongate-lanceolate with simple pinnae (divisions of frond) . . . . **Christmas fern** (*Polystichum acrostichoides*), p. 165

65 Plants upright with stiff branching; fronds broad, tapering, with dissected pinnae . . . . **Bracken fern** (*Pteridium aquilinum*), p. 171

66 Grasses or grasslike plants; leaves alternate and/or basal, mostly elongate, narrow, many times longer than broad; including sedges and rushes, but excluding onions (*Allium* spp.), which are distinguishable by odor and taste; see table 1 (p. 204) for comparisons of the vegetative morphology of grasses, sedges, and rushes 67

66 Herbs (forbs), plants not grasslike; leaves opposite, whorled, alternate, or basal, either simple or compound 123

67 Stems round or flattened in cross section, usually hollow; leaves 2-ranked, from opposite sides of the stem; sheath edges (on side away from leaf blade) characteristically open or overlapping, except closed in *Bromus* (for grasses included in this key); nodes conspicuous; Grass family (Poaceae) 68

67 Stems either more or less angular, usually 3-sided, or round in cross section, mostly solid; leaves 3-ranked, from 3 sides of the stem; sheath closed except in Juncus; nodes inconspicuous 118

68 Stems woody, bamboolike; leaves with short petiole between blade and sheath. **Giant cane** (*Arundinaria gigantea*), p. 31

68   Stems herbaceous; stem leaves without petioles 69

69   Leaves narrow, usually 1/16 in. (1½ mm.) wide or less, sometimes conspicuously curled when dried, with inrolled margins. 70

69   Leaves usually more than 1/16 in. (1½ mm.) wide 76

70   Plants forming small basal clumps consisting of dried curled leaves; leaves widest at base of blade, tapering to fine tip; hairs conspicuous at throat of sheath; plants of thin, dry, or rocky soils. . . . .
. . . . **Poverty oatgrass** (*Danthonia spicata*), p. 221

70   Plants not forming small basal clumps with curling leaves; hairs, if present, not noticeable from general pubescence of blades and sheaths . . 71

71   Perennials forming distinct tufts or clumps 72

71   Perennials or annuals, stems few or solitary 75

72   Leaves mostly basal, either straight and upright or lax and stringlike and forming large arching clumps 73

72   Leaves on stem as well as from ground 74

73   Leaves straight, upright, pubescent or smooth; spikelets in terminal panicles appearing in spring, followed in summer by small clusters hidden among leaves at base of clump . . . .
. . . . **Narrow-leaf panic grass** (*Panicum* spp.), p. 235

73   Leaves stringlike, arching, smooth; spikelets fragrant, in terminal panicles, appearing in late summer . . . .
. . . . **Prairie dropseed** (*Sporobolus heterolepis*), p. 251

74   Leaves generally stringlike when dry; vernation (arrangement of young leaves in shoot as determined by cutting stem in cross section below uppermost full leaf) folded; inflorescence a dense terminal spike . . . .
. . . . **Junegrass** (*Koeleria pyramidata*), p. 231

74   Leaves flat; vernation rolled; inflorescence terminal, with short dense branches, somewhat interrupted . . . .
. . . . **Wedgegrass** (*Sphenopholis obtusata*), p. 249

75   Spikelets conspicuously 3-awned, exserted, on axillary and terminal branches; annual and perennial plants of thin dry soils . . . .
. . . . **Three-awn** (*Aristida* spp.). p. 210

75   Spikelets awnless, more or less hidden in sheathlike leaves; annual plants of limestone plants usually in damp soil glades or open ground . . . .
. . . . **Baldgrass** (*Sporobolus* spp.), p. 250

76   Stem leaves about ⅝ in. (1½ cm.) wideand usually not more than 7 or 8 times as long, with clasping or cordate bases; perennial grasses with green winter rosettes . . . . **Panic grass** (*Dichanthelium* spp.), p. 235

76   Stem leaves usually much narrower but, if as wide as above, then more elongate, at least 10 times as long as wide; leaf base tapering or rounded rather than clasping; rosettes present or absent 77

77   Leaves about ⅝–¾ in. (1½–2 cm.) wide, up to 8 in. (20 cm.) long, sometimes more 78

77 Leaves mostly narrower, ⅜ in. (1 cm.) or less **81**

78 Spikelets few, conspicuously flattened, about ⅝ in. (1½ cm.) wide, nodding, on lax pedicels; woodland plants . . . .
. . . . **Broadleaf uniola (*Chasmanthium latifolium*)**, p. 217

78 Spikelets numerous, not flattened, in spikes or upright panicles; plants of fields, roadsides, or damp ground. **79**

79 Ligule (small projection on the inside junction of the blade and sheath) a collarlike membrane; flowering stalk consisting of a large terminal panicle with numerous brown spikelets . . . .
. . . . **Johnson grass, Sudan grass (*Sorghum* spp.)**, p. 248

79 Ligule consisting mainly of minute hairs; flowering stalk either spikelike and nodding or consisting of several stiff digitate spikes **80**

80 Spikes nodding, terminal, appearing bristly and perennial plants of thin dry soils . . . . **Nodding foxtail (*Setaria faberii*)**, p. 245

80 Spikes stiff, upright, fingerlike; tall robust plants usually in damp soil. .
. . . . **Gamagrass** (Tripsacum dactyloides), p. 253

81 Grasses with closed sheaths . . . .
. . . . **Bromegrass, cheat, chess (*Bromus* spp.)**, p. 213

81 Grasses with split, overlapping, or open sheaths **82**

82 Auricles (2 small appendages at junction of leaf blade and sheath) present; inflorescence a terminal spike, similar to wheat or barley **83**

82 Auricles absent, sometimes rudimentary in meadow fescue, with open branching inflorescence **84**

83 Spreading rhizomes present, forming a sod; spikelets noticeably flattened along the rachis . . . .
. . . . . . **Quackgrass, wheatgrass (*Elymus, Pascopyrum* spp.)**, p. 225, 234

83 Rhizomes absent, not forming a sod; spikelets not noticeably flattened
. . . . **Wild rye (*Elymus* spp.)**, p. 225

84 Basal parts of vegetative stems strongly flattened, sheaths keeled, forming tussocks; late-flowering native perennials to 6–7 ft. (2 m.) high **85**

84 Basal parts mostly rounded, not strongly keeled; sod or tussock grasses (perennials), or annual plants (keeled in early flowering introduced orchard grass) **86**

85 Basal sheaths pubescent, flattened; ligule (small projection on inside junction of leaf and blade) usually conspicuous, about 1/32 in. (¾ mm.) long . . . . **Bluestem, broomsedge (*Andropogon* spp.)**, p. 207

85 Basal sheaths smooth, glabrous; ligule minute, less than 1/32 in. (¾ mm.) long; late-flowering plants with dark, viscid, open panicle branches
. . . . **Purpletop (*Tridens flavus*)**, p. 252

86 Ligule consisting of hairs, or absent **87**

86 Ligule membranaceous with entire, irregular, or lacerated margin . . **99**

87 Ligule absent; coarse annual plants of low ground or damp soils. . . . . . .
. . . . **Barnyard grass (*Echinochloa* spp.)**, p. 224

87   Ligule consisting of hairs, sometimes minute, less than 1/32 in. (¾ mm.) long **88**

88   Leaves straight, upright, from basal clumps, narrow, generally not over ⅛ in. (3 mm.) wide; flowers in terminal panicles appearing early, followed in summer by small clusters hidden among basal leaves . . . . **Narrow-leaf panic grass (*Panicum* spp.), p. 235**

88   Leaves from stem as well as from base **89**

89   Basal parts of stems somewhat keeled, glabrous . . . . **Purpletop (*Tridens flavus*), p. 252**

89   Basal parts of stems mostly rounded **90**

90   Plants with basal rosettes; flowering early from terminal panicles, and later from axillary clusters of lower leaves; sheaths and blades variously pubescent. . . . . **Panic grass (*Panicum* spp.), p. 235**

90   Basal rosettes lacking; flowering once **91**

91   Plants perennial, forming tufts, or spreading by rhizomes or stolons (surface runners) and forming sod **92**

91   Plants annual with only fibrous roots **96**

92   Sod-forming plants, spreading by aggressive stolons; inflorescence digitate, with short narrow spikes **Bermuda grass (*Cynodon dactylon*), p. 218**

92   Tufted or producing loose sod, stolons lacking **93**

93   Leaf blades narrow, about 3/16 in. (4–5 mm.) wide or less, tapering to stringlike tip. **94**

93   Leaf blades wider, not stringlike **95**

94   Inflorescence consisting of short penduous spikes on a slender stalk; upper sheaths with glandular-dotted pubescence; common plants of dry prairies and glades . . . **Sideoats grama (*Bouteloua curtipendula*), p. 212**

94   Inflorescence either partially enclosed in a spathelike sheath or exserted with spreading panicle branches. . . . **Dropseed (*Sporobolus* spp.), p. 250**

95   Leaf blades lax, up to ⅝ in. (1½ cm.) wide or more; inflorescence consisting of 1 to several beadlike racemes from a central stalk . . . . **Paspalum (*Paspalum* spp.), p. 238**

95   Leaf blades firm, about ⅜ in. (1 cm.) wide; ligule conspicuous, about ⅛ in. (3 mm.) long; inflorescence a panicle with stiff straight branches **Switchgrass (*Panicum virgatum*), p. 236**

96   Leaf blades narrow, ⅛ in. (3 mm.) wide or less, often tapering to stringlike tip; short grasses of limestone glades, dry woods, and waste areas . . . . **Baldgrass (*Sporobolus* spp.), p. 250**

96   Leaf blades wider, not stringlike **97**

97   Plants conspicuously pubescent on sheaths and leaves; inflorescence a bushy open panicle . . . . **Witchgrass (*Panicum capillare*), p. 236**

97   Plants generally smooth or with ciliate hairs on sheath margins; inflorescence a bristly spike or upright panicle **98**

110 Ligule conspicuous, usually at least ⅛ in. (3 mm.) long **111**
110 Ligule shorter or reduced to a fringe of hairs **114**

111 Ligule thick, stiff; coarse species to 6–7 ft. (2 m.) tall; of glades and prairies . . . . **Indian grass (*Sorghastrum nutans*)**, p. 247
111 Ligule thin, membranous, whitish; introduced forage plants **112**

112 Stems with bulbous base; hay and pasture grass forming clumps; flower stalk cylindrical . . . . **Timothy (*Phleum pratense*)**, p. 240
112 Stems lacking bulbous base **113**

113 Rhizomes present; sod-forming grass with open somewhat reddish panicle . . . . **Redtop (*Agrostis stolonifera*)**, p. 205
113 Rhizomes lacking; bunch-forming grass with branching inflorescence of dense spikelet clusters or glomerules . . . .
. . . . **Orchard grass (*Dactylis glomerata*)**, p. 220

114 Rhizomes generally present; ligule sometimes reduced to a fringe of hairs; grass of dry woodlands and low ground . . . .
. . . . **Muhly (*Muhlenbergia* spp.)**, p. 232
114 Rhizomes absent **115**

115 Annuals; inflorescence digitate, consisting of several spikelike racemes at top of flower stalk . . . . **Crabgrass (*Digitaria* spp.)**, p. 222
115 Perennials **116**

116 Ligule lacerated, split, whitish; erect narrow-tufted grass. . . . .
. . . . **Wedgegrass (*Sphenopholis* spp.)**, p. 249
116 Ligule entire, whitish or brownish **117**

117 Ligule brown; inflorescence 2 to several beadlike racemes on upper flower stalk . . . . **Paspalum (*Paspalum* spp.)**, p. 238
117 Ligule white; inflorescence a branching panicle . . . .
. . . . **Fescue (*Festuca* spp.)**, p. 227

118 Stems round in cross section, generally leafless or with leaves at base of plant; inflorescence usually subtended by several narrow leaflike bracts . . . . **Rush (*Juncus* spp.)**, p. 230
118 Stems mostly angular or 3-sided, sometimes round in cross section as in the soft-stemmed bulrush; stem leaves, if present, 3-ranked; sheaths closed; plants of the Sedge family (Cyperaceae) **119**

119 Stems leafless; plants of marshy ground or shallow water, forming colonies **120**
119 Stems with leaves; plants of dry or damp ground, not forming colonies **121**

120 Stems triangular, to about 3 ft. (90 cm.) tall . . . .
. . . . **Three-square bulrush (*Schoenoplectus americanus*)**, p. 243
120 Stems rounded, soft, taller, as much as 6–7 ft. (2 m.) . . . .
. . . . **Great bulrush (*Schoenoplectus tabernaemontani*)**, p. 243

121 Coarse plants 5–6 ft. (1.5–1.8 m.) tall, forming tufts; occurring in swales, ditches, and low ground . . . .
. . . . **Bulrush (*Scirpus* spp., *Schoenoplectus* spp.), p. 243**
121 Plants shorter, to 20–25 in. (50–60 cm.) tall; of dry or damp soils . . **122**

122 Inflorescence consisting of narrow spikes or capitate clusters; seed surrounded by a papery saclike structure (perigynium) . . . .
. . . . **Sedge (*Carex* spp.), p. 215**
122 Inflorescence yellowish green, consisting of umbrellalike spikes, subtended by several long narrow leaves and terminating a stiff upright stalk. . . . . **Nutgrass (*Cyperus* spp.), p. 219**

123 Plants with milky juice in leaves and stem. **124**
123 Plants lacking milky juice **130**

124 Leaves opposite or whorled **125**
124 Leaves alternate **127**

125 Flowers inconspicuous; fruit 3-lobed, nutlike, with 3 seeds . . . .
. . . . **Spurge (*Euphorbia* spp.), p. 133**
125 Flowers showy, white, pink, greenish, or orange; fruit a pod with numerous fluffy wind-borne seeds **126**

126 Pods smooth, narrow, pencillilce, usually joined at base . . . .
. . . . **Dogbane (*Apocynum* spp.), p. 100**
126 Pods sometimes rough or hairy, thicker, usually ½ in. (1.2 cm.) or more, but if narrower, than leaves linear, 3–6 at node (leaves alternate in butterflower weed) . . . . **Milkweed (*Asclepias* spp.), p. 103**

127 Fruit 3-lobed, nutlike, with 3 seeds . . . . **Spurge (*Euphorbia* spp.), p. 133**
127 Fruit not as above **128**

128 Flowers deep blue, whitish, or red, scattered along stem; plants of damp or low ground. . . . . **Lobelia (*Lobelia* spp.), p. 148**
128 Flowers bluish or yellow, small, forming more or less radiate heads; plants of glades, prairies, woodlands, or waste ground **129**

129 Leaves ovate to obovate, smooth-margined, with conspicuous pubescence; flowers yellow . . . . **Hawkweed (*Hieracium* spp.), p. 142**
129 Leaves variably dissected, cleft, or coarsely dentate, lacking conspicuous pubescence, upper leaves sometimes entire. . . . .
. . . . **Wild lettuce (*Lactuca* spp.), p. 143**

130 Leaves simple, either entire, toothed, lobed, parted, or coarsely dissected
. . . . **131**
130 Leaves compound, with 3 or more distinct leaflets or finely dissected with numerous divisions **205**

131 Leaves alternate or from base of plant, or both **132**
131 Leaves opposite or whorled, some plants with alternate leaves on upper stem **186**

132 Plants with only 1 or 2 basal leaves; low forest herbs **133**

132   Plants with 3 or more leaves on stem or from base of plant, or both **134**

133   Plants with a single leaf, rounded and somewhat lobed, from the ground; rootstock thick, with reddish juice; flowers white. . . . .
. . . . **Bloodroot (*Sanguinaria canadensis*)**, p. 180

133   Plants with 2 leaves cordate; root aromatic; flowers brownish purple. . .
. . . . **Wild ginger (*Asarum canadense*)**, p. 102

134   Plants with sheathing stipules (ocreae) encircling the stem at the base of the leaf petiole; flowers small; plants of the dock or smartweed family (Polygonaceae) **135**

134   Plants lacking sheathing stipules **136**

135   Plants with mostly simple stems or with few branches; basal rosette usually present; flowers greenish or brownish red**Dock (*Rumex* spp.)**, p. 178

135   Plants with branching stems, lacking basal rosette; flowers pink, white, or greenish white . . . . **Smartweed (*Polygonum* spp.)**, p. 159

136   Leaves mostly smooth-edged, without distinct serrations; teeth, or lobes
. . . . **137**

136   Leaves mostly serrate, toothed, lobed, dissected, parted, or with spiny margins (upper leaves sometimes smooth-edged) **154**

137   Stems and undersides of leaves, whitish with fine matted hairs; leaves linear to ovate-oblong or spatulate and tapering to base; flowers whitish
. . . . **138**

137   Stems and leaves not as above, sometimes with grayish-brown, russet, or whitish pubescence (not matted) **139**

138   Plants with basal rosette of spatulate leaves; flowers in dense terminal clusters . . . . **Pussytoes (*Antennaria plantaginifolia*)**, p. 98

138   Plants lacking rosette; flowers in axillary clusters or, if terminal, then flower stalks branched or more diffuse . . . .
. . . . **Cudweed, everlasting (*Pseudognaphalium* spp.)**, p. 168

139   Leaves all from base of plant, with prominent parallel venation; flower stalks leafless, with narrow greenish or brownish spike . . . .
. . . . **Buckhorn, plantain, ribgrass (*Plantago* spp.)**, p. 164

139   Leaves on stem, although basal Leaves may be present; flowers of various colors but in some species greenish or brown **140**

140   Plants smooth, with thick reddish stems to 6–7 ft. (2 m.) tall or more; leaves at least 2 in. (5 cm.) wide; flowers whitish in large drooping racemes
. . . . **Pokeweed (*Phytolacca americana*)**, p. 163

140   Plants mostly smaller; flowers not as above **141**

141   Leaves mostly stalked **142**

141   Leaves mostly sessile or clasping, at least on upper stem **146**

142   Flowers inconspicuous, greenish or brownish **143**

142   Flowers showy, of other colors **145**

143   Foliage mostly glabrous; leaves oblong-lanceolate; flowers greenish, in spathelike bracts . . . . **Three-seeded mercury (*Acalypha gracilens*)**, p. 88

143 Foliage scurfy to somewhat woolly; leaves linear, ovate, or oblong; plants strongly scented **144**

144 Undersides of leaves whitish to greenish gray; side veins noticeable.... .... **Croton (*Croton* spp.), p. 115**

144 Undersides of leaves distinctly silvery; side veins not noticeable....... .... **Rushfoil (*Croton willdenowii*), p. 117**

145 Flowers solitary, yellow with dark purplish-brown center............. .... **Ground cherry (*Physalis pubescens*), p. 162**

145 Flowers in radiate heads with blue or white rays and yellow disk (center) flowers...................... **Aster (*Symphyotrichum* spp.), p. 191**

146 Plants densely woolly, from a rosette, to 6–7 ft. (2 m.) tall; upper leaves tapering to base and continuing down stem as winged margins; flowers yellow, in dense terminal spikes .. **Mullein (*Verbascum thapsus*), p. 196**

146 Plants not as above **147**

147 Leaves smooth, elliptic-lanceolate with parallel veins, arranged more or less in 2 rows along stem; flowers small, whitish, in terminal clusters .. .. **False Solomon's seal (*Maianthemum racemosum*), p. 150**

147 Leaves without parallel veins and not arranged in 2 rows; flowers in radiate heads, either all yellow, roseate, or purple, sometimes white, or a combination of yellow rays and purplish-brown disk (center) flowers, or white or blue ray flowers and yellow disk flowers; all species of the Aster family (Asteraceae) **148**

148 Leaves linear, 1/16–⅛ in. (1–3 mm.) wide; ray flowers white, blue, or yellow **149**

148 Leaves generally wider, but if sometimes linear and narrow as above, flower heads mostly pink or purple, otherwise ray flowers yellow ... **151**

149 Flowers in radiate heads, all yellow, with drooping rays and globose centers; plants of dry pastures and waste ground........................ .... **Bitterweed (*Helenium amarum*), p. 138**

149 Flowers in radiate heads either all yellow but rays short, not drooping, or with white or blue ray flowers and yellow disk (center) flowers... **150**

150 Flower heads all yellow, in more or less compact to open branching panicles .... **Goldenrod (*Solidago* spp.), p. 185**

150 Flower heads with white or blue ray flowers and generally yellow disk (center) flowers .... **Aster (*Symphyotrichum* spp.), p. 191**

151 Flower heads with yellow rays **152**

151 Flower heads pink, purple, or roseate **153**

152 Flower heads with dark brown disk (center) flowers .... **Black-eyed Susan, coneflower (*Rudbeckia* spp.), p. 175**

152 Flower heads with usually yellow disk flowers .... **Sunflower (*Helianthus* spp.), p. 139**

153 Flower heads pink purplish or roseate, rarely white, in terminal spikes or racemes .... **Blazing star, gayfeather (*Liatris* spp.), p. 146**

153 Flower heads purple, in more or less open or branching inflorescence; leaves sometimes finely serrate...... **Ironweed** (*Vernonia* spp.), p. 198

154 Leaves elongate, with parallel venation and spiny margins; flowers cream colored, in globose heads .... **Rattlesnake master, yucca-leaf eryngo** .... (*Eryngium yuccifolium*), p. 132

154 Leaves not strap-shaped or spiny **155**

155 Flowers with knob or short spur, blue, purplish, sometimes mixed with white, or yellow or cream colored .... **Violet** (*Viola* spp.), p. 200

155 Flowers not as above **156**

156 Leaves only from base of the plant, 3-lobed, with hairy petioles; flowers (or calyx) whitish, pink, or bluish, with 3 lobes..................... .... **Hepatica** (*Anemone americana*), p. 97

156 Leaves present on stem **157**

157 Leaves on stem incised, cleft, or parted, basal leaves (if present) divided or merely toothed **158**

157 All leaves serrate, dentate, or with shallow lobing **162**

158 Leaves with palmate lobing; flowers mostly blue and spurred, or yellow and lacking spur and with 5 distinct petals **159**

158 Leaves with pinnate lobing, some cut almost to midrib, or, if 3-cleft, flowers not as above; flower heads with yellow rays; plants of Aster family (Asteraceae) **160**

159 Flowers blue with conspicuous spur. **Larkspur** (*Delphinium* spp.), p. 122

159 Flowers yellow, not spurred, with 5 distinct shiny petals .... **Buttercup** (*Ranunculus* spp.), p. 173

160 Plants usually less than 3 ft. (90 cm.) tall; ft. heads less than 1 in. (2½ cm.) across........................ **Ragwort** (*Packera* spp.), p. 156

160 Plants generally taller; ft. heads larger, up to 3 in. (7½ cm.) across .. **161**

161 Leaves mostly large, pinnately cleft; plant rough-bristly, of prairies and glades................. **Compass plant** (*Silphium laciniatum*), p. 181

161 Leaves pinnately cleft, or sometimes palmately 3-lobed; plant glabrous, of shaded woodlands and moist soils .... **Coneflower** (*Rudbeckia laciniata*), p. 175

162 Stem leaves as well as basal leaves (if present) with distinct petioles at least ⅛–3/16 in. (3–4 mm.) long **163**

162 Stem leaves mostly sessile, or with elongate tapering base; lower leaves on stem sometimes with distinct petioles; species of Aster family (Asteraceae) **173**

163 Flowers small, inconspicuous, greenish or reddish **164**

163 Flowers showy, white, bluish, or yellow, or in radiate heads with white, blue, pink, purple, or yellow rays **168**

164 The larger leaves on stem at least 1 in. (2½ cm.) wide **165**

164 No leaves more than 1 in. (2½ cm.) wide **166**

175 Leaves ovate-cordate, large, 12 in. (30 cm.) long or more, with long petiole, mostly from base of plant; flower heads yellow . . . . **Prairie dock (*Silphium terebinthinaceum*)**, p. 181

175 Leaves smaller; stems generally leafy **176**

176 Stem leaves tapering to base of blade; basal leaves, if present, also tapering **177**

176 Stem leaves somewhat rounded, clasping, cordate, or sagittate; basal Leaves rounded, cordate, or tapering, sometimes petioled **181**

177 Leaves elongate, finely serrate, with veins running more or less length of blade; flower heads with long drooping yellow or purplish rays. . . . . **Coneflower (*Echinacea* spp.)**, p. 127

177 Leaves not as above; rays not drooping **178**

178 Leaves linear-lanceolate, with long-tapering tip; flower heads small, all purple . . . . **Ironweed (*Vernonia* spp.)**, p. 198

178 Leaves ovate, elliptic, linear, or ovate-lanceolate; flower heads all yellow or white, or with white or pink rays and yellow centers **179**

179 Tall plants, up to 6–7 ft. (2 m.); flowers white; plants of woodlands and waste ground, sometimes abundant after fire. . . . . **Burnweed (*Erechtites hieracifolia*)**, p. 129

179 Plants generally smaller; flower heads all yellow or white, or pink with yellow centers. **180**

180 Flower heads all yellow; plants late flowering . . . . **Goldenrod (*Solidago* spp.)**, p. 185

180 Flower heads white or pink with yellow centers; plants early flowering in most species (see also Symphyotrichum spp.). . . . . **Fleabane (*Erigeron* spp.)**, p. 130

181 Plants rough-hairy; basal rosette usually present; flower heads with yellow rays and purplish-brown globose centers. . . . . **Black-eyed Susan, coneflower (*Rudbeckia* spp.)**, p. 175

181 Plants not as above **182**

182 Basal leaves thick, firm, ovate-lanceolate with one distinct main vein; flower heads whitish, small, about ¼ in. (6–7 mm.) in diameter . . . . **American feverfew (*Parthenium* spp.)**, p. 157

182 Basal leaves not as above; flower heads not all white, generally wider, but if not wider then flowers yellow **183**

183 Flower heads all yellow **184**

183 Flower heads usually with white, pink, or blue ray flowers **185**

184 Leaves thick, harsh; flower heads mostly solitary with conspicuous spreading ray flowers, about 2 in. (5 cm.) across . . . . **Rosin-weed (*Silphium asteriscus*)**, p. 181

184 Leaves not thick; flower heads small, numerous, in dense or spreading panicles. . . . . **Goldenrod (*Solidago* spp.)**, p. 185

185 Flower heads with white or pink flowers, in several whorls, flowering in spring and summer . . . . **Fleabane (*Erigeron* spp.), p. 130**

185 Flower heads with white, blue, or purple ray flowers in a single whorl; late-flowering plants . . . . **Aster (*Symphyotrichum* spp.), p. 191**

186 Leaves whorled, 3 or more at each node, sometimes opposite; flowers small with 4 similar petals; slender, angle-stemmed plants . . . .
. . . . **Bedstraw, cleavers (*Galium* spp.), p. 136**

186 Leaves mostly opposite, 2 at each node **187**

187 Stem square in cross section **188**
187 Stem rounded in cross section **193**

188 Leaves linear, ⅛–3/16 in. (3–5 mm.) wide, 1–2 in. (2½–5 cm.) long . . .
. . . . **189**

188 Leaves usually wider **190**

189 Stipules present, mostly on upper stem; plants turning black upon drying . . . . **Houstonia (*Stenaria nigricans*), p. 172, 188**

189 Stipules lacking; plants with minty odor, not blackening . . . .
. . . . **Mountain mint (*Pycnanthemum tenuifolium*), p. 172**

190 Leaves perfoliate, forming cup around stem; flower heads with yellow rays; tall coarse plant . . . . **Cup plant (*Silphium perfoliatum*), p. 181**

190 Leaves not as above; flowers whitish or purplish to pink or red **191**

191 Leaves mostly sessile, ovate, pointed, about ¾ in. (2 cm.) wide, 1½ in. (4 cm.) long, dotted, with conspicuous side veins; flowers pale, purplish . .
. . . . **Dittany (*Cunila origanoides*), p. 118**

191 Leaves not as above, if ovate-pointed then usually larger, sometimes petioled **192**

192 Plants strongly aromatic; flowers in rounded heads. . . . .
. . . . **Horsemint, bergamot (*Monarda* spp.), p. 154**

192 Plants not particularly odorous; flowers in short purplish spikes . . . . . .
. . . . **Self-heal (*Prunella vulgaris*), p. 167**

193 Leaves dissected or deeply 3-lobed **194**
193 Leaves entire, serrate, or toothed **196**

194 Leaves deeply 3-lobed; flowers in heads with yellow rays . . . .
. . . . **Tickseed (*Coreopsis* spp.), p. 113**

194 Leaves dissected or palmately divided **195**

195 Flowers small, greenish, mostly inconspicuous; plants of fields, pastures, and waste ground . . . . **Ragweed (*Ambrosia* spp.), p. 93**

195 Flowers large, about 2 in. (5 cm.) long, yellow, showy; plant of dry, rocky woods . . . . **Gerardia (*Aureolaria grandiflora*), p. 106**

196 Leaves with entire margins, sometimes with small lateral lobes but not distinctly serrate or toothed **197**

196 Leaves serrate or toothed on the margin **202**

197 Flowers funnelform, blue, pink, or white **198**

197 Flowers in terminal heads with yellow rays, or yellow with 5 similar petals in open spray-like clusters **200**

198 Leaves linear, stiff; flowers small, ¼ in. (6 mm.) long, white, in the axils of the leaves . . . . **Buttonweed** (*Diodella teres*), p. 126

198 Leaves ovate or oblong to linear-lanceolate **199**

199 Leaves ovate-oblong; flowers bluish, resembling petunia, in axils of leaves . . . . **Wild petunia** (*Ruellia* spp.), p. 177

199 Leaves narrow-ovate or linear-lanceolate; flowers blue, pink, or white, in terminal or axillary clusters . . . . **Phlox** (*Phlox* spp.), p. 161

200 Leaves thin, elliptic-oblong, with translucent or black dots, flowers with 5 equal petals . . . . **St. John's-wort** (*Hypericum* spp.), p. 40

200 Leaves not as above **201**

201 Leaves harsh, ovate-lanceolate, occasionally with faint serrations; ray flowers narrow and rounded at tip . . . . **Rosin-weed** (*Silphium integrifolium*), p. 181

201 Leaves glabrous, elongate and narrow, sometimes with small lateral lobes near base of blade; ray flowers broad with toothed tip. . . . . **Tickseed** (*Coreopsis lanceolata*), p. 113

202 Stem leaves sessile, somewhat cordate or clasping; flowers tubular, in loose terminal panicles . . . . **Beardtongue** (*Penstemon* spp.), p. 158

202 Stem leaves either sessile or stalked, sometimes cordate at base of leaf; flowers in small white heads or in broad radiate heads with yellow rays; plants of Aster family (Asteraceae) **203**

203 Leaves thin-textured, petioled; flowers small, white; plant of low or damp woods. . . . . **White snakeroot** (*Ageratina altissima*), p. 90

203 Leaves firm, harsh, or scabrous; flowers yellow, in radiate heads; plants of dry woods and glades **204**

204 Leaves with lateral veins more or less well defined from base of blade and extending upward; ray flowers surrounding a flat receptacle observed when disk (center) flowers removed **Sunflower** (*Helianthus* spp.), p. 139

204 Leaves lacking venation as above; receptacle conical . . . . **Ox-eye** (*Heliopsis helianthoides*), p. 141

205 Leaves usually with 3–9 leaflets **206**

205 Leaves always with more than 10 leaflets, or leaves lacy, finely dissected with small divisions **225**

206 Leaves large, up to 20 in. (50 cm.) long; leaflets 3–5 or more; flowers on a club-shaped spadix enclosed in a spathe; root tuberous with sharp, acrid taste . . . . **Indian turnip, Jack-in-the pulpit, green dragon** . . . . (*Arisaema* spp.), p. 101

206 Leaves smaller, generally less than 6 in. (15 cm.) long **207**

207 Leaflets 3, lacking stipules (small bracts at base of each leaf stalk); low plants with sour-tasting leaves . . . . **Wood sorrel** (*Oxalis* spp.), p. 155

207 Leaflets 3–9; stipules present though sometimes inconspicuous; plants not sour-tasting **208**

208 Leaflets 5 (sometimes 7–9) with conspicuously or coarsely serrate margins; stems spreading or upright; flowers yellow, cup-shaped, with 5 equal petals .... **Cinquefoil (*Potentilla* spp.), p. 166**

208 Leaflets smooth-edged or with minute teeth; flowers white, yellow, cream-colored, pink, roseate-purplish, or bluish, petals not all similar; fruit a pod with 1 to several seeds; plants of Pea family (Fabaceae) **209**

209 Leaflets 3, serrate at least in upper part, sometimes minutely so ... **210**

209 Leaflets 3 or more, mostly smooth-edged, sometimes tip extended as a minute point **212**

210 Tall coarse plants, to 3 ft. (90 cm.) or more; flowers white or yellow, in slender elongate racemes......... **Sweet clover (*Melilotus* spp.), p. 152**

210 Plants mostly shorter, sometimes decumbent; flowers white, yellow, roseate, or purple in rounded or shortened heads................ **211**

211 Stems somewhat angular or square in cross section; lateral leaflets. with short stalks; flowers yellow or purple; fruits curved or coiled.........
.... **Alfalfa, black medick (*Medicago* spp.), p. 151**

211 Stems mostly rounded; lateral leaflets usually sessile; flowers yellow, white, or roseate .... **Clover (*Trifolium* spp.), p. 195**

212 Stems twining, trailing, or reclining **213**

212 Stems upright or stifflower erect **219**

213 Leaflets mostly 5–7 (sometimes 3); stems twining; rootstocks tuberous; pods 2–4 in. (5–10 mm.) long; plants of moist ground ....
.... **Groundnut (*Apios americana*), p. 99**

213 Leaflets typically 3 .... **214**

214 Leaflets more or less rounded or somewhat elliptic, not much longer than broad, with blunt or round apex **215**

214 Leaflets ovate or somewhat triangular or, if elliptic, then several times longer than broad **217**

215 Stems low or reclining; fruit an elliptic, flat, 1-seeded pod...........
.... **Lespedeza (*Lespedeza* spp.), p. 144**

215 Stems twining or creeping; fruit an elongate several-seeded pod ... **216**

216 Leaflets more or less rounded, 1–1½ in (2½–4 cm.) wide, the terminal one sometimes larger than the laterals; stem mostly pubescent; fruit breaking into sections . **Tick trefoil (*Desmodium rotundifolium*), p. 123**

216 Leaflets oblong, all about the same size; fruit a continuous pod, not breaking into sections........... **Milk pea (*Galactia volubilis*), p. 135**

217 Flowers showy, pale blue, 2 in. (5 cm.) long; pods up to 2 in. (5 cm.) long with long-pointed tip...... **Butterflower pea (*Clitoria mariana*), p. 112**

217 Flowers less than 1 in. (2½ cm.) long; pods not long-pointed **218**

218 Pods flat, about ¾ in. (2 cm.) long, from short stalks; plant of damp woods or low ground ...... **Hog peanut (*Amphicarpa bracteata*), p. 96**

218 Pods thickened, up to 2 in. (5 cm.) long, from stalks 2 in. (5 cm.) long or more; plant of dry soils . . . . **Wild bean (*Strophostyles* spp.), p 189**

219 Leaflets 5–7 (sometimes 3), small, linear, mostly less than 3/16 in. (4–5 mm.) wide, about 1 in. (2½ cm.) long; flowers white or roseate, in dense spikes . . . . **Prairie clover (*Dalea* spp.), p. 119**

219 Leaflets usually 3 (excluding large foliaceous bracts or stipules in *Baptisia leucophaea*), or if 5, the leaflets larger and flowers bluish **220**

220 Leaflets with 2 minute needlelike stipules at base; fruit breaking into sections, adhering to clothing or hair . . . .
. . . . **Tick trefoil (*Desmodium* spp.), p. 123**

220 Leaflets not stipellate **221**

221 Leaflets about 1 in. (2½ cm.) long, hairy, with conspicuous side veins, the apex with fine tip; flowers yellow . . . .
. . . . **Pencil flower (*Stylosanthes biflora*), p. 190**

221 Leaflets not as above **222**

222 Plants heavy, thick-stemmed, turning black upon drying; flowers yellow, whitish, or blue; fruit a several-seeded pod, also turning black . . . .
. . . . **False indigo (*Baptisia* spp.), p. 107**

222 Plants not as above; fruit 1-seeded. **223**

223 Leaflets 3; stipules needlelike . . . . . **Lespedeza (*Lespedeza* spp.), p. 144**

223 Leaflets 3 or sometimes 5, stipules broader and more or less bractlike .
. . . . **224**

224 Leaflets 3; low-growing annuals; two introduced species occurring in fields and waste ground . . . .
. . . . **Lespedeza (*Lespedeza stipulacea* and *L. striata*), p. 145**

224 Leaflets 3 or 5; native perennials up to 3 ft. (90 cm.) tall . . . .
. . . . **Prairie turnip, Sampson's snakeroot, scurf-pea**
. . . . (former *Psoralea* spp.), p. 169

225 Leaves with 10 or more distinct leaflets **226**

225 Leaves lacy, finely dissected or divided **232**

226 Leaflets serrate; stems glabrous with purplish mottling, about 1 in. (2½ cm.) thick at base; flowers white in umbellate or flat-topped clusters; toxic plants of low or moist ground growing from a cluster of tuberlike roots. **Water hemlock (*Cicuta maculata*), p. 110**

226 Leaflets entire; plants of Pea family (Fabaceae) **227**

227 Plants thorny and spreading; leaflets about 1/16 in. (1½ mm.) wide; flowers pink or purple in ball-shaped clusters; fruit a prickly pod 1–3 in. (3–7 cm.) long . . . . **Nuttall's mimosa (*Mimosa nuttallii*), p. 153**

227 Plants smooth; leaflets. larger **228**

228 Leaves ending in tendril; flowers white and pale bluish . . . .
. . . . **Wood vetch (*Vicia caroliniana*), p. 199**

228 Leaves lacking tendrils **229**

229 Leaves and stems soft-pubescent, greenish gray **230**
229 Leaves and stems more or less smooth, without grayish cast **231**
230 Leaflets elliptic, with tapering symmetrical base and aristate tip; flowers yellow-purple- pink combination . . . .
. . . . Goat's rue (*Tephrosia virginiana*), p. 194
230 Leaflets oblong, with rounded or cordate somewhat asymmetrical base; flowers purple in terminal fingerlike spikes. . . . .
. . . . Lead plant (*Amorpha canescens*), p. 95
231 Leaflets elliptic-oblong, symmetrical, with terminal leaflet; flowers whitish or blue; fruit a short plump 2-celled pod . . . .
. . . . Milk vetch, ground plum (*Astragalus* spp.), p. 105
231 Leaflets oblong, with irregular base, no terminal leaflet; flowers yellow; fruit a linear 1-celled pod . . . .
. . . . Partridge pea (*Chamaecrista fasciculata*), p. 108

232 Leaves finely dissected, all from base of plant; flowers light pink or whitish; low forest herbs growing from a shallow cluster of small tubers or bulbs . . . . Dutchman's breeches, squirrel corn (*Dicentra* spp.), p. 125
232 Leaves finely dissected, on the stem and sometimes from the base of plant **233**

233 Leaves carrotlike, the base of the petiole U-shaped and clasping the pubescent stem; flowers small, white, in flat-topped clusters; plant of dry ground growing from a taproot . . . .
. . . . Queen Anne's lace (*Daucus carota*), p. 121
233 Leaves and petiole not as above **234**

234 Leaves opposite or sometimes alternate, dissected or deeply lobed, a basal rosette; flowers small, greenish, in terminal spikes or in axils of leaves . . . . Common ragweed (*Ambrosia artemisiifolia*), p. 93
234 Leaves alternate, finely dissected or lobed, also from base of plant or forming rosette; flowers white or blue **235**

235 Leaves finely dissected, elongate; flowers white, in more or less flat-topped clusters . . . . Yarrow (*Achillea millefolium*), p. 89
235 Leaves deeply cut, palmately lobed; flowers mostly blue, in elongate racemes . . . . Larkspur (*Delphinium* spp.), p. 122

WINTER KEY

1   Plants woody; all trees, most shrubs and vines 2
1   Plants herbaceous, usually with wintergreen foliage, sometimes forming basal rosettes; broad-leaved herbs, grasses, and grasslike plants **71**

2   Trees evergreen 3
2   Trees, shrubs, or vines; leaves deciduous......................... 4

3   Leaves small, scalelike or awl-shaped .... **Cedar** (*Juniperus* spp.), p. 44
3   Leaves needlelike, 3 in. (7½ cm.) long or more, in bundles of 2 or 3 ....
    .... **Shortleaf pine** (*Pinus echinata*), p. 55

4   Buds opposite each other, sometimes in whorls of 3 .... **5**
4   Buds alternate **19**

5   Forest trees, sometimes small understory species **6**
5   Low to tall shrubs, or vines **10**

6   Bundle scars 3; twigs mostly slender **7**
6   Bundle scars generally more than 3; twigs stout **9**

7   Buds somewhat elongate, scurfy or fuzzy and brown or rust colored, or smooth and grayish; fruit berrylike, black or bluish black, 1-seeded....
    .... **Blackhaw** (*Viburnum* spp.), p. 84
7   Buds smooth, light brown, greenish, or tawny; fruit 2-seeded **8**

8   Fruit ovoid, berrylike, red, white, blue, or bluish black, 2-seeded ......
    .... **Dogwood** (*Cornus* spp.), p. 28
8   Fruit a double-winged samara .... **Maple** (*Acer* spp.), p. 13

9   Lateral buds pointed, with several overlapping light brown or rose-hued scales............................. **Buckeye** (*Aesculus* spp.), p. 15
9   Lateral buds more or less obtuse, reddish brown, brown, or black, dull or somewhat scurfy .... **Ash** (*Fraxinus* spp.), p. 35

10 Shrubby plants with scaly or stringy bark, light brown to deep reddish brown... **11**
10 Shrubs or vines, no shedding bark **14**

11 Young twigs brownish to deep reddish brown; low shrubs 1–3 ft. (30–90 cm.) tall **12**
11 Young twigs pale tan or straw colored; tall shrubs, to 6 ft. (2 m.) high ..
    .... **13**

12 Bark deep reddish brown; fruit a dry capsule; bushy shrub of pine and oak woods......... **St. Andrew's Cross** (*Hypericum hypericoides*), p. 42
12 Bark brownish, somewhat stringy; fruit a small reddish berry in axillary clusters, persistent in winter; clumping shrub of prairies, open woods, and pastures .... **Coralberry** (*Symphoricarpos orbiculatus*), p. 78

13 Young twigs angular in cross section; leaf scar minute, inconspicuous; bushy shrub, densely branched . **St. John's-wort** (*Hypericum* spp.), p. 40
13 Young twigs rounded; leaf scar somewhat V-shaped, with 3 bundle scars; shrub sparsely branched **Wild hydrangea** (*Hydrangea arborescens*), p. 39

| 14 | Bundle scar 1; shrubs .... **15** |
| 14 | Bundle scars 3 or more .... **17** |

15 Leaf scars almost circular, surrounding the bud; fruit dry, in a globular aggregate about ¾ in. (2 cm.) in diameter, persistent in winter; plants of swamps and wet areas .. **Buttonbush (*Cephalanthus occidentalis*)**, p. 26

15 Leaf scars semicircular or shallow, if circular then below the bud and not surrounding it **16**

16 Twigs greenish or reddish, mostly glabrous, sometimes with slight ridges; lateral buds with smooth scales; fruit reddish in winter .... **Eastern wahoo (*Euonymus atropurpureus*)**, p. 34

16 Twigs brownish with scurfy pubescence, lacking ridges; lateral buds with brownish tomentum; fruit purplish to bluish pink, spherical, very small and numerous, in clusters**French mulberry (*Callicarpa americana*)**, p. 19

17 Bundle scars more than 3; twigs mostly stout; usually shrublike .... **Red Buckeye (*Aesculus pavia*)**, p. 15

17 Bundle scars 3 .... **18**

18 Lenticels large, raised; lf. scars large; pith white; conspicuous coarse shrub .... **Elderberry (*Sambucus nigra*)**, p. 73

18 Lenticels not conspicuous; leaf scars small; the pith not as above; bark scaly or stringy .... **Honeysuckle (*Lonicera* spp.)**, p. 47

19 Twigs and branches spiny, prickly, or thorny, or with short pointed spurs **20**

19 Twigs without spines, prickles, or thorns **30**

20 Milky sap present; inner bark orange; tree of fence rows, pastures, open ground.................... **Osage-orange (*Maclura pomifera*)**, p. 48

20 No milky sap; trees, shrubs, or vines **21**

21 Climbing or shrubby vines with green stems and coiling tendrils .... **Greenbrier (*Smilax* spp.)**, p. 76

21 Trees or shrubs, if trailing or climbing then lacking tendrils; bundle scars 3 .... **22**

22 Shrubs, upright or trailing **23**

22 Trees solitary, or small and forming thickets **25**

23 Prickles needlelike; bark sometimes scaly; upright forest shrubs .... **Wild gooseberry (*Ribes* spp.)**, p. 66

23 Prickles with broad bases; shrubs, upright or trailing **24**

24 Leaf scar narrow, a smooth pencillike line; stems upright, mostly greenish or reddish .... **Rose (*Rosa* spp.)**, p. 69

24 Leaf scar not smooth; an irregular break; stems trailing or consisting of heavy upright canes, greenish, bluish white, or reddish .... **Dewberry, blackberry, raspberry (*Rubus* spp.)**, p. 71

25 Stipules spiny, one on each side of the leaf scar; buds obscure beneath leaf scar (the Prickly-ash, *Zanthoxylum americanum*, has a distinctly noticeable bud .... **Black locust (*Robinia pseudoacacia*)**, p. 68

37 Bundle scars numerous; lf. scar oval; buds brownish, somewhat fuzzy; small shrub of dry upland sites, with pungent-tasting twigs . . . .
. . . . **Aromatic sumac (*Rhus aromatica*),** p. 63

37 Bundle scars 3; tall shrubs or trees with spicy-flavored twigs **38**

38 Twigs greenish or greenish yellow, smooth; fruit blue, 1-seeded . . . . . .
. . . . **Sassafras (*Sassafras albidum*),** p. 74

38 Twigs brownish, smooth; fruit red, 1-seeded . . . .
. . . . **Spicebush (*Lindera benzoin*),** p. 45

39 Trees and shrubs with milky sap on fresh cuts **40**

39 Trees and shrubs without milky sap **41**

40 Leaf scar almost encircling buds; twigs smooth, thick; fruit dry, reddish, in dense terminal clusters, persistent; shrub, forming colonies . . . .
. . . . **Smooth sumac (*Rhus glabra*)** p. 63

40 Leaf scar hemispherical, below bud; tree . . . .
. . . . **Red mulberry (*Morus rubra*),** p. 51

41 Shrubs **42**

41 Small or large trees **50**

42 Twigs grayish brown, sometimes with 2 lateral buds above the leafscar, arranged one above the other; tall shrub to 10 ft. (3 m.) or more; of low ground and creekbanks . . . . **False indigo (*Amorpha fruticosa*),** p. 95

42 Twigs not as above **43**

43 Aerial roots present; twigs mostly pubescent, yellowish brown; buds downy tan or yellowish lacking overlapping scales; bundle scars more than 3 . . . . **Poison ivy (*Toxicodendron radicans*),** p. 63

43 Aerial roots lacking; twigs not as above. **44**

44 Bark peeling; young branches reddish, older ones somewhat grayish; bundle scars 3, distinct . . . **Missouri currant (*Ribes missouriense*),** p. 66

44 Bark not peeling; bundle scars 1 or 3 . . . . **45**

45 Buds with scales, mostly smooth **46**

45 Buds downy or scurfy, lacking scales, if scales present then distinctly fuzzy as seen with hand lens **48**

46 Bundle scars 1; low or tall shrubs on generally acid soils. . . . .
. . . . **Blueberry, deerberry, sparkleberry (*Vaccinium* spp.),** p. 82

46 Bundle scars 3; mostly tall shrubs **47**

47 Twigs grayish or light brown; buds mostly pointed; species of limestone soils . . . . **Lance-leaved buckthorn (*Rhamnus lanceolata*),** p. 62

47 Twigs reddish or brownish, with gland-tipped hairs as viewed with hand lens; buds oval, blunt . . . . **Hazelnut (*Corylus americana*),** p. 30

48 Bundle scars 1; buds fuzzy; low shrubs less than 3 ft. (1 m.) tall, with slender upright branches and dry persistent panicles . . . .
. . . . **New Jersey tea, redroot (*Ceanothus* spp.),** p. 23

48 Bundle scars 3; buds downy, appearing naked; shrubs mostly taller. . **49**

49 Lateral buds elongate, somewhat stalked in appearance, yellowish brown; fruit a 2-celled woody capsule .... **Witch-hazel (*Hamamelis* spp.)**, p. 38

49 Lateral buds not stalked, brownish; fruit bluish black, spherical, ¼ in. (6 mm.) in diameter, 3-seeded, somewhat persistent..................
.... **Carolina buckthorn (*Rhamnus caroliniana*)**, p. 62

50 Twigs with conspicuous brownish or chocolate-colored chambered pith; leaf scar large .... **Butternut, walnut (*Juglans* spp.)** p. 43

50 Twigs lacking chambered pith **51**

51 Buds with glossy or varnished appearance, resinous, with several overlapping scales; bundle scars 3; twigs sometimes with corky wings or ridges; fruit globular, dry, persistent ....
.... **Sweetgum (*Liquidambar styraciflua*)**, p. 46

51 Buds not as above; bundle scars 1, 3, 5, or more **52**

52 Terminal buds narrow, elongate; bundle scars 3 or more; small trees **53**

52 Terminal buds not as above **54**

53 Buds smooth, with several brown or pinkish overlapping scales; bundle scars 3; plants of dry upland woods and bluffs ....
.... **Juneberry, serviceberry, shadbush (*Amelanchier arborea*)**, p. 16

53 Buds hairy, dark rusty brown, without overlapping scales; bundle scars 5 or more; plants of moist ravines and bottom lands.................
.... **Pawpaw (*Asimina triloba*)**, p. 17

54 Twigs heavy, with large heart-shaped leaf-scars, sometimes with more than 1 small flat greenish lateral bud (one above the other); bundle scars mostly 3–5; pith salmon colored; pods thick, woody, about 5 in. (12 cm.) long, and persistent . **Kentucky coffeetree (*Gymnocladus dioicus*)**, p. 37

54 Twigs not as above, if thick then with large terminal bud and numerous bundle scars; pith not pink **55**

55 Bundle scars 1 .... **56**

55 Bundle scars 3 or more **57**

56 Buds to ⅛ in. (3 mm.) long, obtuse, with 2 bud scales; bark on trunk thick, fissured, in squarrose blocks on mature trees ....
.... **Persimmon (*Diospyros virginiana*)**, p. 33

56 Buds much smaller, with 3 or more overlapping scales; bark not fissured or in squarrose blocks ...... **Sparkleberry (*Vaccinium arboreum*)**, p. 82

57 Pith saffron or yellowish brown and conspicuous; twigs reddish brown or purplish, smooth, with whitish lenticels; small tree of limestone glades and barrens........... **American smoketree (*Cotinus obovatus*)** p. 31

57 Pith not as above; small or large forest trees of various habitats **58**

58 Buds rusty brown, tomentose, blunt or rounded; twigs stout, dark, sometimes with short spurlike twigs; fruit black, berrylike; generally a small tree of dry slopes and glades ....
.... **Gum bumelia (*Sideroxylon lanuginosum*)**, p. 75

58 Buds and twigs not as above **59**

59 Generally small trees at maturity, in the understory, sometimes in open ground; bundle scars 3 or appearing as 3 .... **60**

59 Large forest trees; bundle scars 3 or more **64**

60 Twigs generally with a single bud at the tip similar to the lateral buds **61**

60 Twigs with a true terminal bud generally larger than the lateral buds, or several buds clustered together at the tip **63**

61 Buds minute; twigs dark, characteristically zig-zag; old bark somewhat furrowed and plated; fruit a short flat pod, persistent in winter .... **Redbud (*Cercis canadensis*)**, p. 27

61 Buds conspicuous; twigs olive brown or reddish brown; fruit not a pod **62**

62 Bundle scars 3; buds smooth, pointed, with striated scales as seen with a hand lens; bark scaly; fruit a papery capsule...................... .... **Eastern hophornbeam (*Ostrya virginiana*)**, p. 53

62 Bundle scars more, appearing in 3 groups; buds dull, ovoid, lacking striated scales; bark in longitudinal strips; fruit a spiny bur............. .... **Ozark chinkapin (*Castanea ozarkensis*)**, p. 22

63 Buds ovoid with minute scales, somewhat fuzzy; twigs reddish brown (short spurs may be present)......... **Wild crab (*Malus ioensis*)**, p. 49

63 Buds elongate, naked, without scales; twigs tawny; spurs never present .... **Carolina buckthorn (*Rhamnus caroliniana*)**, p. 62

64 Twigs generally with a single bud at the tip similar to the lateral buds; bundle scars 3 .... **65**

64 Twigs with a true terminal bud generally larger than the lateral buds, or several buds clustered together at the tip; bundle scars 3 or more ... **69**

65 Twigs and branches with smooth, thin, bitter-tasting, and ill-smelling bark; lenticels conspicuous, elongate .... **Black cherry (*Prunus serotina*)**, p. 56

65 Twigs and branches not as above; lenticels not conspicuous **66**

66 Buds with 2 unequal scales, smooth, reddish, and plump............. .... **Basswood (*Tilia americana*)**, p. 79

66 Buds not as above, if smooth then not reddish **67**

67 Buds triangular and appressed against twig; twigs frequently with "witches' brooms;" older bark warty...... **Hackberry (*Celtis* spp.)**, p. 25

67 Buds not triangular or appressed against twig; bark not warty **68**

68 Twigs with numerous small lenticels; buds with mostly 2 visible scales; bark in flattened longitudinal light-colored strips .... **Ozark chinkapin (*Castanea ozarkensis*)**, p. 22

68 Twigs not as above; buds with 5 or more visible scales; bark fissured in irregular strips .... **Elm (*Ulmus* spp.)**, p. 80

69 Bundle scars 3, distinct; pith white, in vertical section with delicate cross walls....................... **Blackgum (*Nyssa sylvatica*)**, p. 52

69 Bundle scars more than 3; pith lacking cross walls **70**

70 Terminal buds large, scaly or sometimes clavate (2 opposing scales of equal size); leaf scar heart-shaped or triangular, conspicuous . . . . **Hickory (*Carya* spp.), p. 20**

70 Terminal buds blunt or pointed, usually several clustered at end of the twig, with numerous small overlapping scales; leaf scar shallow or semicircular; pith stellate in cross section . . . . **Oak (*Quercus* spp.), p. 58**

71 Stems leafless, tubular, with green jointed sections, sometimes with whorls of leafless branches . . . . **Horsetail (*Equisetum* spp.), p. 128**

71 Stems not leafless and tubular **72**

72 Fern with low-arching or prostrate elongate fronds, usually dark green in winter . . . . **Christmas fern (*Polystichum acrostichoides*), p. 165**

72 Grasses, grasslike plants, or broad-leaved herbs (wildflowers, forbs) . **73**

73 Grasses or grasslike plants **74**

73 Broad-leaved herbs **78**

74 Stems leafy, mostly triangular in cross section, without nodes or joints; leaves green to bluish green or glaucous, sometimes thick textured; inflorescence a loose to compact cluster of saclike bracts, each with a single seed . . . . **Sedge (*Carex* spp.), p. 215**

74 Stems rounded or oval in cross section, with or without nodes **75**

75 Stems slender, wiry, usually tough, without nodes; leaves stringlike; inflorescence usually subtended by several stringy bracts . . . . **Rush (*Juncus* spp.), p. 230**

75 Stems with nodes; leaves not stringlike; true grass species **76**

76 Coarse grass forming dense clumps; leaveselongate, somewhat shiny, and dark green with conspicuous venation; auricles sometimes present with cilia on margin. . . . **Tall fescue (*Schedonorus arundinaceus*), p. 227**

76 Grasses forming sod or, if tufted, not as above **77**

77 Rhizomes present, sod-forming; leaf blades generally not over ½ in. (3–4 mm.) wide; ligule collarlike . . . . **Bluegrass (*Poa* spp.), p. 241**

77 Basal rosettes present, not sod-forming; leaf blades varying from ½ in. (3–4 mm.) to 1 in. (2½ cm.) wide; ligule, if present, consisting of hairs . . . . **Panic grass (*Panicum* spp., *Dicanthelium* spp.), p. 235**

78 Stems and leaves usually with conspicuous mat of whitish pubescence **79**

78 Stem and leaves smooth or, if hairy or pubescent, not a whitish mat **80**

79 Leaves linear or spatulate, forming a noticeable basal rosette . . . . **Pussytoes (*Antennaria* spp.), p. 98**

79 Leaves on stem, not forming exclusively a rosette . . . . **Cudweed (*Pseudognaphalium* spp.) p. 168**

92   Leaves broadly triangular, cordate, or arrow-shaped, to narrow oblong or lanceolate; venation mostly conspicuous. . **Violet** (*Viola* spp.), p. 200

92   Leaves more or less rounded or reniform (occasionally some basal leaves may be divided), somewhat fleshly or firm, smooth, with blunt rounded teeth . . . . **Buttercup** (*Ranunculus* spp.), p. 173

93   Leaves oval, elliptic, or cordate, smooth, dentate with numerous teeth, mostly purplish beneath . . . . **Ragwort** (*Packera* spp.), p. 156

93   Leaves usually broadly lanceolate to cordate-sagittate, serrations pointed, sometimes remote, or margins entire, in some species purplish beneath . . . . **Aster** (*Symphyotrichum* spp.), p. 191

94   Basal leaves entire or remotely serrate, thin to coarse-textured . . . . **Aster** (*Symphyotrichum* spp.), p. 191

94   Basal leaves mostly serrate, rough or coarse, or if entire then mostly coarse, with conspicuous venation . . . **Goldenrod** (*Solidago* spp.), p. 185

WINTER KEY TO SELECTED OAKS (*QUERCUS*)

1   Buds rounded or egg-shaped, with more or less blunt apex, reddish brown; acorns maturing in 1 year, present on the current season's growth **2**

1   Buds acute or pointed, light brown to reddish; acorns maturing in 1 year as above, or in 2 years and immature on current twigs and fully developed on wood of the preceding year **4**

2   Buds mostly smooth; twigs slender, somewhat purplish to reddish purple; acorns about 1 in. long (2½ cm.), shiny, the cup scales warty; leaves, if persistent, with several narrow rounded lobes and deep sinuses. . . . . **White oak (*Quercus alba*), p. 58**

2   Buds somewhat pubescent; acorns ovoid. **3**

3   Buds light brown, to ¼ in. (6 mm.) long; acorn large, 1 in. (2½ cm.) long or more, with deep cup enclosing ½ or more of the nut, with conspicuous fringe at margin; leaves obovate with shallow lobing in upper part, tapered at base . . . . **Bur oak (*Quercus macrocarpa*), p. 58**

3   Buds reddish brown, about ⅛ in. (3 mm) long; acorns smaller, only about ½ in. (1.2 cm.) long, the cup enclosing about ⅓ of the nut, not fringed; leaves thick, cross-shaped, with wide truncated lobes. . . . . **Post oak (*Quercus stellata*), p. 58**

4   Twigs brown or grayish: acorns oblong, maturing in I season, the cup with thin scales; leaves oblong-obovate to oblanceolate with coarse-toothed margins and parallel side veins . . . . **Chinkapin oak (*Quercus muehlenbergii*), p. 58**

4   Twigs mostly reddish brown, sometimes lighter colored or with grayish sheen over darker bark; acorns maturing in 2 seasons; leaves variable, elliptic, entire-margined to deeply lobed, with spiny tips **5**

5   Buds more or less smooth, at least the lower scales. **6**

5   Buds pubescent, somewhat scurfy, or with fine tomentum **9**

6   Buds dull, yellowish brown, glabrous; acorns with flat, saucer-shaped cup; leaves with numerous bristly tips and usually deep rounded sinuses . . . . **Shumard oak (*Quercus shumardii*), p. 58**

6   Buds light reddish brown to deep red. **7**

7   Terminal buds dull, egg-shaped, about 3/16 in. (5 mm.) long or less; acorns mostly spherical with cup covering about ½ of the nut; leaves elliptic, without lobing. . . . . **Shingle oak (*Quercus imbricaria*), p. 58**

7   Terminal buds mostly shiny or glabrous; acorns spherical to oblong . . **8**

8   Terminal buds about ¼ in. (6 mm.) long or less; acorns small, spherical, with vertical lines, the cap flat, shallow; leaves spiny-tipped, with 5–7 lobes, the lower lobes somewhat horizontal . . . .
. . . . **Pin oak** (*Quercus palustris*), p. 58

8   Terminal buds generally larger, to ⅜ in. (9 mm.) or even more; acorns oblong, to 1 in. (2½ cm.) long with mostly shallow cup; leaves spiny-tipped with 7 or more lobes, more oblique . . . .
. . . . **Northern red oak** (*Quercus rubra*), p. 58

9   Terminal buds about ¼ in. (6 mm) long or less; twigs reddish brown, slender 10

9   Terminal buds larger; twigs grayish brown to somewhat reddish or black, stout 12

10   Twigs light reddish brown to brownish; acorns mostly spherical; leaves elliptic, without lobing . . . . **Shingle oak** (*Quercus imbricaria*), p. 58

10   Twigs reddish brown to dark brown 11

11   Buds somewhat obtuse, with grayish tomentum toward tip; acorns large, ovoid-oblong, up to 1 in. (2½ cm.) long, with a deep cup, sometimes with a series of circular markings around tip; leaves with 7 or more spiny-tipped lobes and deep egg-shaped sinuses. . . . .
. . . . **Scarlet oak** (*Quercus coccinea*), p. 58

11   Buds pointed, with reddish or brownish pubescence; acorns smaller, somewhat spherical, about ½ in. (1.2 cm.) long, with a shallow cup; leaves with several relatively simple lobes, sometimes with spiny tips, and a narrowing inverted-bell-shaped base . . . .
. . . . **Southern red oak** (*Quercus falcata*), p. 58

12   Buds with grayish pubescence; acorns large with heavy cup, the scales more or less free at tip; leaves obovate with broad spiny-tipped lobing and shallow sinuses; inner bark yellow to orange . . . .
. . . . **Black oak** (*Quercus velutina*), p. 58

12   Buds with reddish brown pubescence; acorns medium-sized, scales broad, somewhat truncated; leaves obovate, with 3 shallow lobes slightly indented, with wedge-shaped base . . . .
. . . . **Blackjack oak** (*Quercus marilandica*), p. 58

Glossary

Acaulescent. Lacking a stem, as for some plants like squirrel corn.

Achene. Small, one-seeded dry fruit, as in the sunflower family.

Acicular. Needlelike.

Acuminate. Tapering gradually to a point or apex as for leaf blades.

Acute. Terminating in a sharp point.

Alternate. Arrangement of leaves on the stem, with one leaf at each node.

Ament. Small inflorescence, usually a pendulous spike, with either male or female flowers, as in willows, or with only male flowers as in oaks, hickories, and several other tree genera; same as catkin.

Annual. Completing growth and reproduction in one season, after which the plant dies. A winter annual germinates in the fall and completes its cycle the following spring.

Arborescent. Treelike in size and general appearance.

Aristate. Bristle-tipped or awned, on several grasses.

Attenuate. Tapering narrowly and gradually; a drawn-out tip.

Auricle. Minute appendage or lobe located at the summit of the sheath for some grasses and rushes.

Awl-shaped. Slender, sharp-pointed.

Awn. Slender bristle of various lengths.

Axil. Angle formed by the stem and the leaf (or between any two structures or organs).

Axis. Vertical or longitudinal portion, as for an inflorescence.

Berry. Soft or juicy fruit with several seeds.

Bipinnate. Doubly pinnate, with reference to compound twice-divided leaves.

Bract. Usually small foliaceous or awl-shaped organ subtending flowers or flower clusters; sometimes the reduced leaves on the upper stem of some plants.

Calyx. Outer whorl of mostly green parts below the petals of a flower.

Capitate. Headlike, usually dense, with reference to an inflorescence of numerous small individual flowers as in species of the sunflower flowers and various other plants.

Capsule. Many-seeded dry fruit which splits when ripened as in Jimson weed and many other plants.

Catkin. Same as ament.

Ciliate. With fringe of hairs on the margin.

Cleft. Cut or incised, like some leaf blades.

Compound leaf. A leaf with blade divided into several distinct leaflets.

Connate. United, joined together.

Cordate. Heart-shaped, referring to a leaf blade.

Corolla. The whorl of petals of a flower, separate or united.

Crenate. With rounded teeth on margin or base of leaf blade.

Crown. Branching portion of the tree or shrub; junction of stem and root in a herbaceous plant.

Culm. The stem of grasses or sedges.

Decreaser. A species that decreases in population density or cover with continuous overgrazing.

Decumbent. Prostrate or touching the ground, as applied to the stem.

Dehiscent. Splitting open at maturity.

Dentate. Toothed, as on the margin of the leaf blade.

Digitate. Fingerlike, referring to arrangement of spikes at the summit of a flower stalk, or to leaflets in some leaves.

Dioecious. Separate male and female flowers, each on different plants as in cottonwood.

Disk. The center part of a composite head, surrounded by ray flowers, or collectively referring to disk flowers as opposed to ray flowers.

Dissected. Cut or divided into narrow lobes, a leaf blade with deep incisions.

Downy. With soft fine hairs.

Drupe. Soft fruit with one stony pit, as in cherries and plums.

Elliptical. Widest at the middle and tapering similarly to both base and apex.

Entire. With no teeth or serrations on margins.

Exserted. Extending out of the sheath as a grass inflorescence.

Falcate. Curving to the tip, sickle-shaped.

Fascicle. Cluster, bundle.

Filiform. Threadlike.

Floret. The part of the spikelet comprising the lemma, palea, and flower.

Foliar. Referring to foliage or leaves.

Frond. Leaflike part of the fern plant.

Glabrous. Smooth, no hairs or roughness.

Glade. A dry, shallow soil area with sparse vegetation, and sometimes small trees and shrubs. Valuable chiefly for grazing and wildlife food and cover. Glades occupy over a million acres in the White River drainage area of southwestern Missouri and northern Arkansas.

Glaucous. With a whitish bloom.

Glumes. Pair of empty (sterile) bracts or scales below the grass floret(s).

Habit. General shape of a plant.

Habitat. Natural location or site of a plant or animal, or of a community of plants and animals.

Hastate. Arrow-shaped.

Head. Dense cluster of flowers.

Herbaceous. Without perennial or woody stem.

Hirsute. With usually coarse hairs.

Hispid. With bristly hairs.

Incised. Deeply cut.

Increaser. A plant already present in an area which increases in abundance under overgrazing.

Inflorescence. Flowering stalk or cluster of flowers.

Internode. Section of the stem or culm between two successive nodes.

Involucre. Bracts surrounding a flower or flower cluster, as in sunflower.

Irregular. Unequal or dissimilar parts, with reference to the corolla of a flower.

Keeled. Sharply folded or with a dorsal ridge as a sheath.

Lacerate. With irregular indentations or cuts.

Lanceolate. Usually elongate and tapering gradually to the apex.

Leaflet. Separate leaflike division of a compound leaf.

Legume. Pod, characteristic of the Pea family, splitting along both edges.

Lemma. One bract of the grass floret, opposite the palea.

Lenticel. Warty spots on bark of stems and branches as in elderberry, similar in function to stomata.

Ligule. Collarlike or hairy projection of grasses at the junction of the sheath and leaf blade; also used in reference to strap-shaped ray flowers in the Aster family.

Linear. Narrow, with parallel sides.

Lobe. Segment of a leaf rounded or angular at the apex, as for many oak leaves.

Loment. Special pod of the Pea family which breaks up into one-seeded parts as in the tick trefoils.

Lyrate. Broadest in the upper part of the leaf with a rounded apex, and incised and lobed toward narrowing base, as in bur oak.

Membranaceous. Thin, papery, or translucent.

Mebranous. Same as membranaceous.

Midrib. The main or central vein of a leaf.

Monoecious. Separate male and female flowers on the same plant as in oak, hazelnut, and hickory.

Node. Joint or location on stem from which leaves arise.

Nut. Hard, dry, 1-seeded fruit such as an acorn, hickorynut, or hazelnut.

Obcordate. Inverted cordate, with broadest part toward the apex.

Oblanceolate. Inverted lanceolate, with broadest part toward the apex.

Oblique. Unequal or asymmetrical as for the base of leaf blade of elm.

Oblong. Longer than broad, with rounded apex and base.

Obovate. Inverted ovate, with broadest part toward the apex.

Obtuse. Blunt or wide-angled, as an apex of a leaf.

Ocrea. Sheathing stipule, encircling the stem, as in dock and smartweed.

Ocreae. Plural of ocrea.

Opaque. Dark, not translucent.

Orbicular. Circular in outline.

Oval. Rounded-elliptic.

Ovary. Organ of the flower containing seeds.

Ovate. Egg-shaped in outline.

Palea. One bract of the grass floret, opposite the lemma and usually smaller than the latter.

Palmate. Lobes or veins arising from base of leaf blade.

Palmately compound. Leaflets all from summit of petiole or leaf stalk, as in buckeye.

Panicle. A type of compound inflorescence, much-branched, with stalked flowers or spikelets.

Papillose. Having hairs or projections with a swollen or glandlike base.

Parted. Deeply cut as in certain leaf blades, usually to the midrib.

Pedicel. Small stalk of a single flower or spikelet in a compound inflorescence.

Peduncle. Principal stalk of the inflorescence or head, also of a single flower where only one normally present.

Pellucid. Transparent.

Peltate leaf. With petiole attached to blade on the lower side away from margin, sometimes at center.

Pendulous. Hanging down, limp.

Perennial. Plant persisting for several years.

Perfect flower. One with both male and female parts.

Perfoliate leaf. Stem passing through the blade of leaf.

Perianth. Whorls of the flower including calyx and corolla, or calyx only if petals absent.

Perigynium. Saclike enclosure for the fruit (achene) of sedges in the genus *Carex*.

Petiole. Stalk of leaf.

Petioled. With a petiole.

Pilose. With soft hairs.

Pinnae. Leaflets or divisions of a compound leaf, or of a fern frond.

Pinnate. Featherlike arrangement of veins, lobes, or leaflets along the midrib or main vein of leaf blade.

Pinnately compound. With leaflets of a compound leaf arranged along the rachis or central axis.

Pinnatifid. Lateral incisions or cuts toward the midrib of leaf blade.

Pistillate. Referring to the female flowers.

Pith. Spongy or central part of the stem or twig.

Pome. Fleshy fruit of the apple type.

Prickle. Short sharp-tipped growth from surface bark or epidermis, as in blackberry.

Pubescent. With hairs.

Raceme. Simple unbranched inflorescence of stalked flowers or spikelets.

Rachis. Central axis of a compound leaf or flowering stalk.

Rays. Strap-shaped flowers around the central disk as in sunflower.

Regular. Equal and similar parts, with reference to petals of a flower.

Reniform. Broadly lobed at base with short blunt apex, as in some leaf blades; kidney-shaped.

Rhizome. Underground stem rooting at the nodes and producing new plants.

Rib. The main vein or veins of the leaf.

Root. Underground organ, lacking nodes.

Rootstock. Rhizome.

Rosette. Basal cluster of leaves.

Runner. Stolon.

Sagittate. Arrow-shaped, similar to hastate but basal lobes not spreading outward.

Samara. Dry winged fruit of elm, ash, and maple.

Scabrous. Rough-surfaced.

Scurfy. Mealy or scaly on the surface.

Segment. Lobes or parts of dissected leaf blade.

Serrate. Sawlike or sharp-toothed on the margin of the leaf, the teeth pointed toward apex.

Sessile. Lacking a stalk or petiole.

Sinus. The notch or space between lobes of a leaf.

Smooth. Glabrous, lacking pubescence; or margins without indentations.

Spadix. The fleshy central floral structure bearing flowers as in Jack-in-the-pulpit.

Spatulate. Spoon-shaped; widest in the upper part, rounded at the apex and with long tapering base.

Spicate. Resembling a spike.

Spike. Simple unbranched inflorescence of sessile flowers or spikelets.

Spikelet. Referring to grasses and sedges, the small individual structure comprising one or several florets; may be either sessile or with a stalk.

Spine. Sharp-pointed growth of variable length from the surface of branches or stem, as in roses or blackberry.

Sporadic. Scattered, erratic distribution.

Staminate. Referring to the male flowers.

Stellate. Star-shaped, as with starlike clusters of pubescence on a leaf blade.

Stigma. Upper part of the style that receives pollen. Stipule. Bractlike appendage on each side of the base of the leaf petiole in certain plants.

Stolon. Prostrate or creeping runner rooting at nodes and producing new plants.

Tendril. Slender clasping stem or appendage, usually coiling.

Thorn. Sharp-pointed reduced woody branch.

Tomentose. Densely pubescent or woolly.

Truncate. Straight across, as if cut off.

Tuber. Thick or rounded underground stem with buds.

Umbel. A type of compound inflorescence, flat-topped or umbrella-shaped, pedicels arising from a common point.

Unisexual. Of single sex, either male or female.

Vernation. The arrangement of the unexpanded leaf blade in the sheath envelope, as in grasses, as seen in bud cross-section.

Vinous. With long soft hairs.

Woolly. With densely matted hairs.

Index by Plant Family

ACANTHACEAE, ACANTHUS FAMILY
Ruellia (Ruellia, Wild Petunia), 177

ALTINGIACEAE, SWEET GUM FAMILY
Liquidambar styraciflua (Sweetgum, Redgum), 46

AMARANTHACEAE, AMARANTH FAMILY
Chenopodium (Lamb's Quarters, Pigweed, Wormseed)), 109

AMARYLLIDACEAE, DAFFODIL FAMILY
Allium (Wild Garlic, Wild Onion), 91

ANACARDIACEAE, SUMAC FAMILY
Cotinus obovatus (American Smoketree), 31
Rhus, Toxicodendron (Sumac, Poison Ivy), 63

ANNONACEAE, CUSTARD-APPLE FAMILY
Asimina triloba (Pawpaw), 17

APIACEAE, CARROT FAMILY
Cicuta maculata (Water Hemlock), 110
Daucus carota (Queen Anne's Lace), 121
Eryngium yuccifolium (Rattlesnake Master), 132

APOCYNACEAE, DOGBANE FAMILY
Apocynum (Dogbane, Indian Hemp), 100
Asclepias (Butterfly-weed, Milkweed), 103

ARACEAE (ARUM FAMILY)
Arisaema triphyllum (Jack-in-the-Pulpit), 101

ARISTOLOCHIACEAE, BIRTHWORT FAMILY
Asarum canadense (Wild Ginger), 102

ASTERACEAE, ASTER FAMILY
Achillea millefolium (Milfoil, Yarrow), 89
Ageratina altissima (White Snakeroot), 90
Ambrosia (Ragweed, Horseweed), 93
Antennaria plantaginifolia (Indian Tobacco, Pussytoes), 98
Coreopsis (Tickseed), 113
Echinacea (Coneflower), 127
Erechtites hieraciifolius (Burnweed), 129
Erigeron (Fleabane, Horseweed), 130
Helenium (Bitterweed, Sneezeweed), 138
Helianthus (Sunflower), 139
Heliopsis helianthoides (Ox-eye), 141

GROSSULARIACEAE, CURRANT FAMILY
Ribes (Currant, Gooseberry), 66

HAMAMELIDACEAE, WITCH-HAZEL FAMILY
Hamamelis vernalis (Ozark Witch-hazel), 38

HYDRANGEACEAE, HYDRANGEA FAMILY
Hydrangea arborescens (Wild Hydrangea), 39

HYPERICACEAE, ST. JOHN'S-WORT FAMILY
Hypericum (St. John's-wort), 40
Hypericum hypericoides (St. Andrew's Cross), 42

JUGLANDACEAE, WALNUT FAMILY
Carya (Hickory), 20
Juglans (Butternut, Walnut), 43

JUNCACEAE, RUSH FAMILY
Juncus (Rush), 230

LAMIACEAE, MINT FAMILY
Cunila origanoides (Dittany), 118
Monarda (Horsemint, Wild Bergamot, Beebalm), 154
Prunella vulgaris (Self-heal, Heal-all), 167
Pycnanthemum tenuifolium (Mountain Mint), 172

LAURACEAE, LAUREL FAMILY
Lindera benzoin (Spicebush), 45
Sassafras albidum (Sassafras), 74

LILIACEAE, LILY FAMILY
Maianthemum racemosum (False Solomon's Seal), 150

MALVACEAE, MALLOW FAMILY
Tilia americana (American Basswood, Linden), 79

MENISPERMACEAE, MOONSEED FAMILY
Menispermum canadense (Moonseed), 50

MORACEAE, MULBERRY FAMILY
Maclura pomifera (Osage-orange), 48
Morus rubra (Red Mulberry), 51

NYSSACEAE, TUPELO FAMILY
Nyssa sylvatica (Blackgum, Black Tupelo), 52

OLEACEAE, ASH FAMILY
Fraxinus (Ash), 35

Scientific Name Index

Common Name Index

www.ingramcontent.com/pod-product-compliance
Lightning Source LLC
Chambersburg PA
CBHW052109030426
42335CB00025B/2899